6th International Conference on Electronic Devices, Systems, and Applications (ICEDSA 2017)

IOP Conference Series: Materials Science and Engineering
Volume 340

Kuching, Malaysia
7 – 8 August 2017

ISBN: 978-1-5108-6116-9
ISSN: 1757-8981

Printed from e-media with permission by:

Curran Associates, Inc.
57 Morehouse Lane
Red Hook, NY 12571

Some format issues inherent in the e-media version may also appear in this print version.

Printed by Curran Associates, Inc. (2018)

For permission requests, please contact the Institute of Physics at the address below.

Institute of Physics
Dirac House, Temple Back
Bristol BS1 6BE UK

Phone: 44 1 17 929 7481
Fax: 44 1 17 920 0979

techtracking@iop.org

Additional copies of this publication are available from:

Curran Associates, Inc.
57 Morehouse Lane
Red Hook, NY 12571 USA
Phone: 845-758-0400
Fax: 845-758-2633
Email: curran@proceedings.com
Web: www.proceedings.com

Table of contents

Volume 340

6th International Conference on Electronic Devices, Systems and Applications 2017 (ICEDSA 2017)

7–8 August 2017, Kuching, Sarawak, Malaysia

Accepted papers received: 16 March 2018
Published online: 9 April 2018

Preface

Papers

IOP Conf. Series: Materials Science and Engineering **340** (2018) 011001 doi:10.1088/1757-899X/340/1/011001

6th International Conference on Electronic Devices, Systems and Applications 2017 (ICEDSA 2017)

[1]**Uzer Mohd Noor** (Chair)

[1]**Abdul Hadi Abdul Razak** (Chair)

[1]**Sukreen Hana Herman** (Treasurer)

[1]**Hashimah Hashim** (Publication)

[1]**Puteri Sarah Mohamad Saad** (TPC I)

[1]**Syed Abdul Mutalib Al Junid** (TPC II)

[1]**Zurita Zulkifli** (Publicity)

[1]**Ahmad Sabirin Zoolfakar** (Local Arrangement)

[2]**Mazlina Mansor Hassan** (Local Arrangement)

[2]**Bibi Sarpinah Sheikh Naimullah** (Local Arrangement)

[1]**Shafinaz Sobihana Shariffudin** (Registration)

[1]Centre for Electronic Engineering Studies, Faculty of Electrical Engineering, Universiti Teknologi MARA, Shah Alam, Selangor, Malaysia

[2] Faculty of Electrical Engineering, Universiti Teknologi MARA Cawangan Sarawak, Kota Samarahan, Sarawak Malaysia

Abstract. On behalf of the Organizing Committee of ICEDSA2017, and with great pleasure, we would like to welcome all delegates to Kuching, Sarawak, Malaysia. Being held from 7th to 8th August 2017 at Hilton Hotel, Kuching, the event is organized by the Centre for Electronic Engineering Studies, Faculty of Electrical Engineering, Universiti Teknologi MARA (UiTM), Malaysia. The conference creates a forum for engineers, scientists and practitioners of varied disciplines and their associated applications. Throughout this event we hope to create an opportunity for researchers, academics and students to get together and, more importantly, to welcome new peers in diverse areas of expertise. On behalf of the organizing committee we would like to take the opportunity to express our gratitude to all reviewers who have been working hard to review the papers on time and share their expertise. Special mention goes to the distinguished Keynote Speakers for their willingness to spend their time and share their valuable knowledge. My deepest gratitude goes to the Organizing Committee who have been working hard behind the scenes for the success of this event. To all participants, we would like to encourage you to fully utilize this platform for knowledge sharing and networking while enjoying your stay in Kuching, Sarawak, Malaysia.

Peer review statement

All papers published in this volume of *IOP Conference Series: Materials Science and Engineering* have been peer reviewed through processes administered by the proceedings Editors. Reviews were conducted by expert referees to the professional and scientific standards expected of a proceedings journal published by IOP Publishing.

IOP Conf. Series: Materials Science and Engineering **340** (2018) 012001 doi:10.1088/1757-899X/340/1/012001

Current-Voltage Characteristics of Nb_2O_5 nanoporous via light illumination

Nur Samihah Khairir, Rozina Abdul Rani, Wan Fazlida Hanim Abdullah, Mohamad Hafiz Mamat, Rosmalini Abdul Kadir, M. Rusop, Ahmad Sabirin Zoolfakar

Faculty of Electrical Engineering, University Teknologi Mara, 40450 Shah Alam, Malaysia

Email: samihahkhairir@gmail.com or ahmad074@salam.uitm.edu.my

Abstract. This work discussed the effect of light on I-V characteristics of anodized niobium pentoxide (Nb_2O_5) which formed nanoporous structure film. The structure was synthesized by anodizing niobium foils in glycerol based solution with 10 wt% supplied by two different voltages, 5V and 10V. The anodized foils that contained Nb_2O_5 film were then annealed to obtain an orthorhombic phase for 30 minutes at 450°C. The metal contact used for I-V testing was platinum (Pt) and it was deposited using thermal evaporator at 30nm thickness. I-V tests were conducted under different condition; dark and illumination to study the effect of light on I-V characteristics of anodized nanoporous Nb_2O_5. Higher anodization voltage and longer anodization time resulted in higher pore dispersion and larger pore size causing the current to increase. The increase of conductivity in I-V behaviour of Nb_2O_5 device is also affected by the illumination test as higher light intensity caused space charge region width to increase, thus making it easier for electron transfer between energy band gap.

1. Introduction

Current technology is in need of a new material breakthrough to catch up into the ever-vast trend of modern devices. One of the potential material that has yet to be fully discovered is Niobium (Nb). It has started to grow in interest with recent years due to its relatively abundant in nature, and interesting structural phases [1]. The oxide form of niobium, Niobium Pentoxide (Nb_2O_5) had been researched as early as 1940s because of its many polymorphic forms and from there the exploration of Nb_2O_5 kickstarted. Its characteristics of having high corrosion resistance, thermodynamically stable with unique performance of the energy band diagram and the identification of crystal phases had made the material to be further investigated [2].

Some of the earliest applications of Nb_2O_5 as functional material were the in the production of sensors, acting as catalyst and used in electrochromism [3]. The first level of Nb_2O_5 research was in either in thin layer, bulk or suspensions but recently the synthesis process of producing various Nb_2O_5 morphology had grown in popularity. Since the advancement of nanotechnology, Nb_2O_5 research had been focused to a more complex study of morphology.

Several synthesis methods have been reported to obtained nano-structured Nb_2O_5, in which some of them are anodization [4,5], sol–gel dip-coatings , hydro-thermal, pulsed laser deposition and electrodeposition [2,3,6–8]. Due to anodization's capability of forming highly porous and ordered oxide morphology to Nb_2O_5, it is one of the most widespread nano-fabrication methods. Anodization method of Nb_2O_5 synthesis is in the most peak interest among others are also due to its advantages of having low fabrication cost and controllable film thickness.

The growth and morphology of the anodic oxide film strongly depends on the applied anodization potential, the composition of the electrolyte, the electrolyte temperature, and the anodization duration. The as-anodized Nb_2O_5 films are mainly amorphous and require a post-annealing treatment to be converted into highly crystalline and stoichiometric Nb_2O_5 [9,10].

For Nb_2O_5 to be applied as devices, there are various ways of fabrication techniques with different atmospheric exposures can be done. Different medium used in the fabrication gives a huge role in contributing to its performance, particularly current-voltage (I-V) characteristics in electronic devices. I-V behaviours can be affected by atmospheric exposures such as temperatures, external currents, radio frequency and also the illumination of light [11–13]. The light presents has an effect on I-V behaviour of the device by changing the width of the space charge region (SCR). This change directly affects the reverse bias I-V curves by increasing the current values and the slope of the curves. The photodetection performance of the devices tends to increase with the increasing light power due to the increasing SCR width [14].

To the extent of author's knowledge, currently there is no research or publication towards Nb_2O_5 I-V characterization to the effect of light illumination. In this paper, the project focused on the I-V behaviour of anodize Nb_2O_5 with different anodization variables on the effect of light illumination. The I-V results showed that there is slight increase of current when samples were being exposed to light.

2. Experimental Procedure

2.1. Fabrication of nanoporous Nb_2O_5

Niobium foils with dimensions of 2.5 cm × 1 cm (99.9% purity, Sigma Aldrich) were gone through ultrasonic cleaning with methanol and deionized water before they were anodized at different time ranging from 2min, 4 min and 6min in different voltage range (5V, 10V and 15V) to produce Nb_2O_5 nanoporous structures. The anodization was conducted using the electrolyte containing 10 wt.% K_2HPO_4 (98% purity, Sigma Aldrich) in anhydrous glycerol (99% purity, Sigma Aldrich), at 180°C as suggested previously from different works. The as-anodized samples were annealed in air at 450°C for 30 min (ramp up/down of 2°C per min). The Nb_2O_5 has many polymorphic forms which gives rise to an interesting series of structural phases.

2.2. Characterization

For structural characterization, Field Emission Scanning Electron Microscopy (FESEM) was conducted to study the surface morphology of the sample using JEOL JSM-7600F. Electrical connections to the Schottky gas sensors were achieved by physically connecting the probes to the deposited Pt metal (top electrode) and Nb foil (bottom electrode). The current–voltage (I–V) measurements were carried out using the two- point probe of Keithley 4200 as they were tested in a dark condition and illuminated from -4V to 4V applied voltage.

3. Results

3.1. FESEM images

The FESEM images in Figure 1 show the surface morphology of the synthesized annealed Nb_2O_5 layer. Looking along the 2-min anodization of Figure 1(a-i), 1(b-i), and 1(c-i), as voltage applied increases from 5V, 10V and 15V, the dispersion of pores appeared to be higher, thus lowering the pore density. The same case is also shown at 6-min anodization of Figure 1(a-ii), 1(b-ii), 1(c-ii). Additionally, if the diameter of pore size is being compared, the diameter size increased as the anodization time increased. At 5V anodization of 2 min, the pore size is at average around 7nm while the at 6 min anodization, the average pore size is at 9nm range.

For 10V anodization of 2 min, the average pore size of nanoporous Nb_2O_5 is around 8nm, while at 6 min the average pore size is larger, which is approximately 13nm, an indication showing higher anodization voltage formed higher pore size. The same case also applied on 15V anodization where at 2-min synthesis, the average pore size is around 10nm while at 6-min synthesis the average pore size is 20nm. Figure 1(d) showed the cross-section image of 6 min anodized Nb_2O_5 at 15V where 'vein-like' nanoporous structure can be seen.

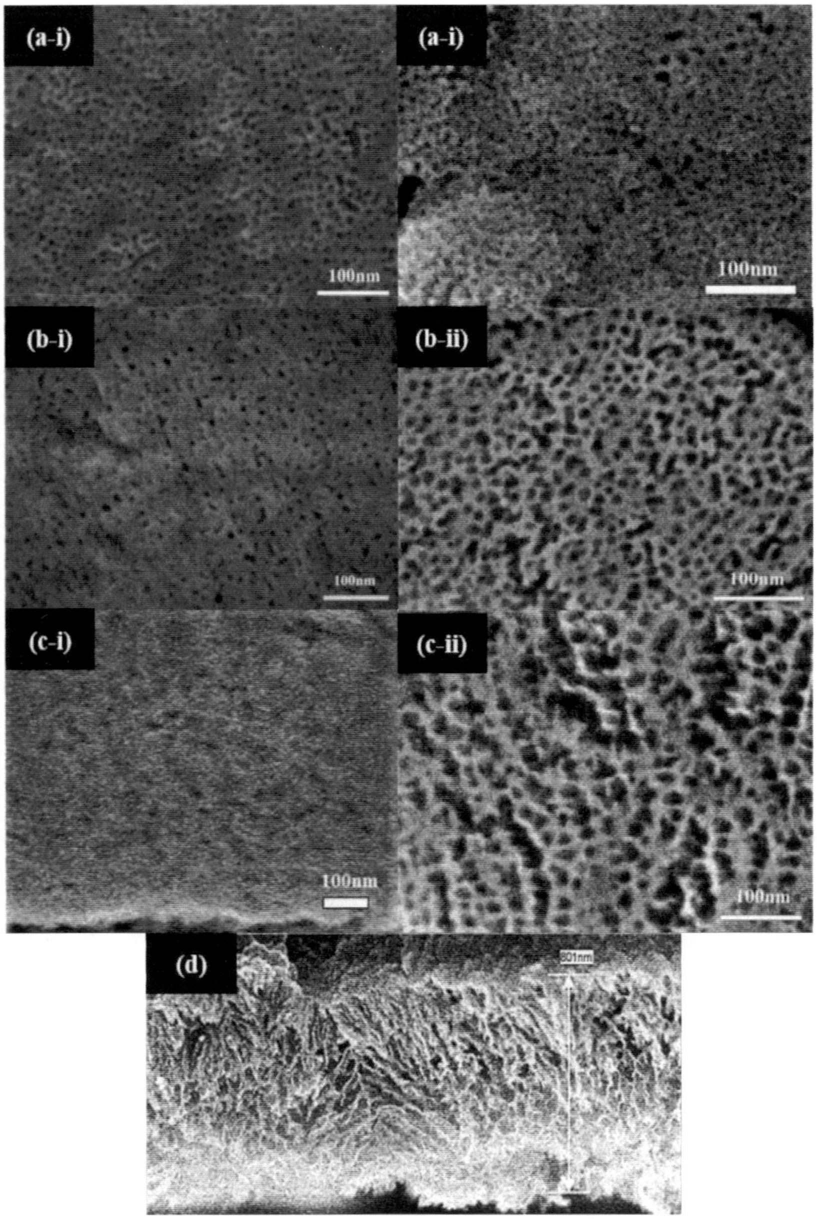

Figure 1. Surface morphology in 5V, 10V, and 15V in (a), (b), and (c) respectively with (i) in 2-min and (ii) in 6-min anodization time, and (d) cross-section image of anodized Nb_2O_5.

3.2. I-V Test

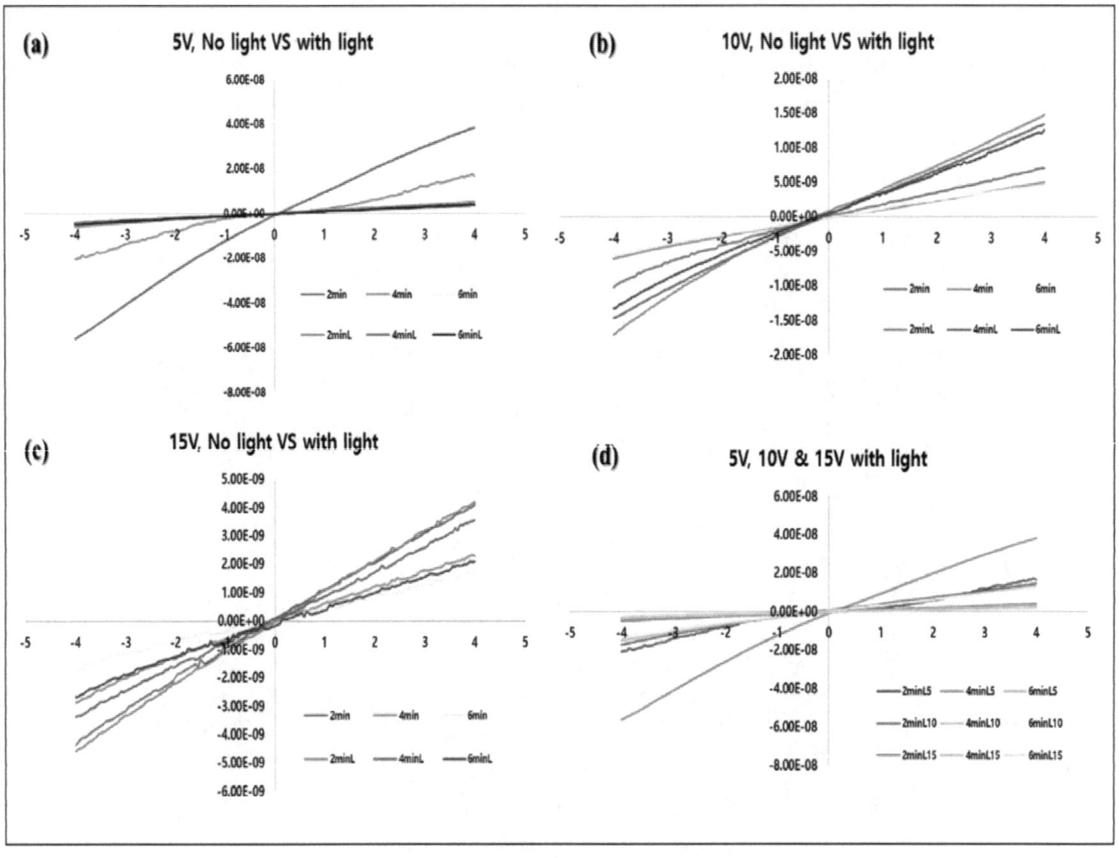

Figure 2. I-V test results comparison of (a) 5V, (b) 10V, and (c) 15V Nb_2O_5 anodization of dark VS illumination test, while (d) focuses on overall I-V illumination test result comparison of 5V, 10V, and 10V Nb_2O_5 anodization.

Figure 2 illustrates a rather similar I-V pattern occurred for Figure 2(a), 2(b) and 2(c). The plots of I-V show preferably linear relation in both dark and light condition, and specifically obtained higher current when the samples were tested under illumination. According to Palatnikov et al [15], light exposure leads to the formation of fractal micro and nanostructures in Nb_2O_5, which have a significant effect on its thermal expansion, thus making it prone to harbor higher energy that caused it to produce higher current.

Other researchers [16–19] also suggested the same case where photon energy from the light makes the band gap of the material decreases, thus making it easier for the charge carrier to crosses over and increase the rate of current density. The increasing light power increases SCR width of the material thus the photodetection performance of devices tends to increase. These factors can contribute to the conductivity of the film since the flow of electrons is moving faster at higher dispersion of pores and bigger pore sizes [19] since it provides continuous and directional paths for electron transfer. When photon energy is present, electrons gain higher energy to jump across the band gap.

The directional paths transfer also affected the I-V characteristic of the device with different thickness. This can be seen when the voltage synthesis of anodized Nb_2O_5 is being compared (Figure 2(d)) in which the result suggested lower anodization voltage produced higher current than that of higher anodization voltage.

CONCLUSION

This study presented the current-voltage characteristics of Nb_2O_5 nanoporous via light illumination. Higher anodization voltage produces bigger pore sizes and larger pore dispersion. Along with longer anodization time, the pore size also increases significantly which leads to encouragement of electron transfer, which in relation produce higher current. The case also follows when I-V is tested on different photon exposure. Compared with those in dark condition, the magnitude of the current is larger under illumination. However, when the current level is being compared between different anodization voltage, lower anodization time produced higher current since it produces thinner layers of Nb_2O_5. The increasing change in light intensity caused the photodetection performance of the device to increase due to the increasing SCR width, thus producing higher jump rate of electrons to move across the band gap.

ACKNOWLEDGEMENT

The authors would like to acknowledge with gratitude the NANO Electronic Centre of Faculty of Electrical Engineering UiTM for the facilities. This work is supported by Ministry of Education Malaysia (MOE) under the Research Assistant Grant Scheme (RAGS) (Project code: 600-RMI/RAGS 5/3 (6/2015)).

REFERENCES

[1] Rani R A, Zoolfakar A S, O'Mullane A P, Austin M W and Kalantar-Zadeh K 2014 Thin films and nanostructures of niobium pentoxide: fundamental properties, synthesis methods and applications *J. Mater. Chem. A* **2** 15683–703

[2] Lee C S, Kwon D, Yoo J E, Lee B G, Choi J and Chung B H 2010 A highly sensitive enzyme-amplified immunosensor based on a nanoporous niobium oxide (Nb2O5) electrode *Sensors* **10** 5160–70

[3] Wen H, Liu Z, Wang J, Yang Q, Li Y and Yu J 2011 Facile synthesis of Nb2O5 nanorod array films and their electrochemical properties *Appl. Surf. Sci.* **257** 10084–8

[4] Abdul Rani R, Zoolfakar A S, Subbiah J, Ou J Z and Kalantar-Zadeh K 2014 Highly ordered anodized Nb2O5 nanochannels for dye-sensitized solar cells *Electrochem. commun.* **40** 20–3

[5] Ghosh R, Brennaman M K, Uher T, Ok M-R, Samulski E T, McNeil L E, Meyer T J and Lopez R 2011 Nanoforest Nb2O5 photoanodes for dye-sensitized solar cells by pulsed laser deposition. *ACS Appl. Mater. Interfaces* **3** 3929–35

[6] Rani R A, Zoolfakar A S, Oua J Z, Field M R, Austin M and Kalantar-Zadeh K 2013 Nanoporous Nb2O5 hydrogen gas sensor *Sensors Actuators, B Chem.* **176** 149–56

[7] Bharti D C and Rhee S-W 2013 Dielectric properties and X-ray photoelectron spectroscopic studies of niobium oxide thin films prepared by direct liquid injection chemical vapor deposition method *Thin Solid Films* **548** 195–201

[8] Zhitomirsky I 1998 Electrolytic deposition of niobium oxide films *Mater. Lett.* **35** 188–93

[9] Kadir R A, Rani R A, Zoolfakar A S, Ou J Z, Shafiei M, Wlodarski W and Kalantar-Zadeh K 2014 Nb2O5 Schottky based ethanol vapour sensors: Effect of metallic catalysts *Sensors Actuators, B Chem.* **202** 74–82

[10] Rani R A, Zoolfakar A S, Ou J Z, Kadir R A, Nili H, Latham K, Sriram S, Bhaskaran M, Zhuiykov S, Kaner R B and Kalantar-zadeh K 2013 Reduced impurity-driven defect states in anodized nanoporous Nb2O5: the possibility of improving performance of photoanodes. *Chem. Commun. (Camb).* **49** 6349–51

[11] Chand S and Kumar J 1997 Effects of barrier height distribution on the behavior of a Schottky diode *J. Appl. Phys.* **82** 5005–10

[12] Rabah K V O 1994 The effects of temperature on schottky diode barrier height and evidence

of multiple barrier *Int. At. Energy Agency Urfited Nations Educ. Sci. Cult. Organ.*

[13] Gupta R K, Ghosh K and Kahol P K 2009 Effect of temperature on current-voltage characteristics of Cu2O/p-Si Schottky diode *Phys. E Low-Dimensional Syst. Nanostructures* **41** 876–8

[14] Shockley W 1949 The Theory of p-n Junctions in Semiconductors and p-n Junction Transistors *Bell Syst. Tech. J.* **28** 435–89

[15] Palatnikov M N, Shcherbina O B, Frolov A A, Efremov V V and Sidorov N V 2011 Effect of High Intensity Light on the Properties and Structure of Ceramic Nb 2 O 5 **47** 619–23

[16] Imer A G and Ocak Y S 2016 Effect of Light Intensity and Temperature on the Current Voltage Characteristics of Al/SY/p-Si Organic–Inorganic Heterojunction *J. Electron. Mater.* **45** 5347–55

[17] Jang S L, Chen Y S and Chang P C 1993 Optical effects on the current-voltage characteristics of lightly doped drain silicon MOSFETs *Solid State Electron.* **36** 727–34

[18] Li J, Wang D and Ba D 2012 E ff ects of Temperature and Light Illumination on the Current – Voltage Characteristics of Molecular Self-Assembled Monolayer Junctions

[19] Rahman M M, Rani R A, Sadek A Z, Zoolfakar A S, Field M R, Ramireddy T, Kalantar-zadeh K and Chen Y 2013 A vein-like nanoporous network of Nb2O5 with a higher lithium intercalation discharge cut-off voltage *J. Mater. Chem. A* **1** 11019

IOP Conf. Series: Materials Science and Engineering **340** (2018) 012002 doi:10.1088/1757-899X/340/1/012002

Effect of substrates on Zinc Oxide thin films fabrication using sol-gel method

Rosmalini Ab Kadir, Nurmalina Mohd Taib, Wan Rosmaria Wan Ahmad, Anees Abdul Aziz, Ahmad Sabirin Zoolfakar.

Faculty of Electrical Engineering, University Teknologi Mara, 40450 Shah Alam, Malaysia

Email: amaltaib56@gmail.com or rosma391@salam.uitm.edu.my

Abstract. The properties of ZnO thin films were deposited on three different substrates via dip coating method was investigated. The films were prepared on glass, ITO and p-type silicon. Characterization of the film revealed that the properties of the dip coated ZnO thin films were influenced by the type of substrates. The grains on ITO and glass were ~10 nm in size while the grains on wafer agglomerate together to form a denser film. Studies of the optical properties using UV-VIS-NIR of the fabricated films demonstrated that glass has the highest transmittance compared to ITO.

1. Introduction

Zinc Oxide (ZnO) is a very important II-VI group of semiconductor material with a direct bandgap of 3.37 eV at room temperature, excellent chemical and thermal stability, non-toxicity and has a large exciton binding energy (60 meV) [1]. Researches on ZnO thin films have gained a lot of attention due to its unique physical properties such as structural, optical and morphological as well as its wide range of applications. ZnO films have been studied for diverse of applications such as optoelectronic devices [2], solar cells [3,7], gas sensors [4], varistors [5], surface acoustic wave devices [6], and field effect transistors [8-10].

ZnO thin films can be prepared by various methods including chemical vapor deposition [11], sputtering [12], pulsed laser deposition [13], sol-gel [14, 15], thermal evaporation [16], anodization [17] and oxidation [18]. However, among these methods, sol-gel dip coating method has several advantages such as low fabrication costs, simple process, large area of substrate coating, allows control over the composition and dopant incorporation. In this work ZnO thin films deposited on different type of substrates which are glass, ITO and silicon wafer by the sol–gel dip-coating technique. The influence of different substrates on the properties of ZnO thin films was investigated. The characterization of the ZnO thin films is performed by field emission scanning electron microscope (FESEM) to study the morphological properties, X-ray diffractometer (XRD) for crystallinity properties and UV-VISS for optical properties.

2. Experimental Procedure

Fabrication of nanoporous Nb_2O_5

In this experiment, zinc acetate dihydrate (Zn.2O) was used as a precursor while 2-methoxyethanol () and monoethanolamine (MEA) (NO) were used as the solvent and the stabilizer respectively. To prepare the ZnO thin films, zinc acetate dihydrate was dissolved in a mixture composed of 2-methoxyethanol and MEA. The zinc acetate dihydrate and MEA molar ratio was kept at 1:1, while the concentration of zinc acetate was 0.4M. The solution was stirred and heated at 80 °C on the hot plate for 1 hour until it become clear and homogeneous. Then, the solution was sonicated at 50 °C for 1 hour by using an ultrasonic water bath. Afterwards, the solution was being stirred without any heat for 24 hours before the deposition process. In this experiment, dip-coating method was used to obtain the ZnO thin films on the substrates.

The glass and ITO substrates were cleaned ultrasonically twice in acetone, methanol and deionized water for 5 minutes before the deposition process. As for silicon substrates, before continuing with the normal cleaning process, we need to soak the substrates with a solution of hydrofluoric acid (HF) and deionized water (DI) with a ratio of 1:10. The substrates were dipped and withdrawn from the precursor solution at a withdrawal speed of 1.0mm/s under ambient conditions. Immediately after the coating process, the thin films were heated to 300 °C for 10 minutes to evaporate the solvent. The final coatings were obtained after repeating this process for five times. After coating, the thin films were annealed for 1 hour at 500 °C under ambient condition.

The structural properties of the ZnO thin films were characterized by using a field emission scanning electron microscope (FESEM) for surface morphology and surface profiler to calculate the thickness of the thin film for calculations of band gap energy. The optical properties were characterized by UV-VIS-NIR to investigate the transmittance and absorption of the thin films.

3. Results

FESEM images Structural properties

Fig. 1 shows the FESEM images of ZnO films fabricated under identical conditions on (a) glass, (b) ITO and (c) silicon. Each transparent substrate has unique surface conditions. Therefore, the deposited ZnO thin films on the different transparent substrates will have different morphology. The SEM images revealed that the fabricated thin films generally consist of grains with varied sizes and dimensions. The ZnO films appeared to have about the same grains size on glass and ITO substrates. However the films fabricated on glass has more voids. The ZnO films become more dense and tend to agglomerate on silicon wafer leaving large grain boundaries. Generally, ZnO thin films on all substrates appeared to be in porous structures.

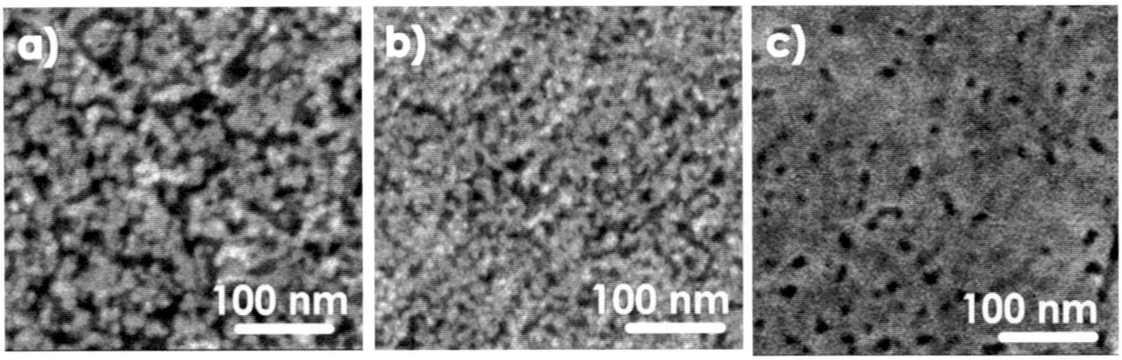

Figure 1. FESEM morphology of ZnO thin films deposited at three different substrates: a) glass, b) ITO and c) silicon.

Opti cal Properties

The optical properties such as transmittance and absorption coefficient were investigated by analyzing the UV-Vis-NIR patterns of ZnO thin films in the wavelength range of 300-800 nm depicted in Fig. 2 and Fig. 3. The transmittance of ZnO thin films on glass substrate increased abruptly from 10%-85% in the visible range (380-800 nm). Furthermore, the optical transmission for ZnO thin films on ITO substrate also increased rapidly in the visible range but only until 80%. It is suggested that the presence of voids around the grains is greater on glass compared to ITO that makes the light to transmit higher on glass, which was previously discussed in the FESEM results. In addition, silicon has 0% optical transmission since it is not transparent substrate.

Many electrons or disordered lattice structures in the ZnO thin films were scattered by incident light, thus decreasing the transparency of ZnO thin films [19]. Furthermore, the difference of optical transmittance between substrates was due to optical scattering at the grain boundaries. It is suggested that the transmittance of the ZnO thin films depends on films structure and uniformity [20]. Fig.3 shows the absorption coefficient of ZnO thin films in three different substrates that were calculated by using Lambert's Law, where α is the absorption coefficient, t is the thin film's thickness and T is the transmittance of the thin film, as shown in Equation 1.

$$\alpha = \frac{1}{t}\ln\left(\frac{1}{T}\right) \tag{1}$$

ITO has larger absorption coefficient compared to glass substrate in the UV region (<400 nm) that might be due to the amount of zinc species within the films and the density of the ZnO thin films on the ITO substrate that can cause an improvement in the light absorption in the visible region. The band gap energy of the thin films has been determined by using the Equation 2 and Equation 3 [22, 23] below:

$$\alpha hv = B(hv - E_g)^{1/2} \tag{2}$$

$$(\alpha hv)^{1/2} = B(hv - E_g) \tag{3}$$

Where α is the absortion coefficient in Fig. 3, hv is the photon energy, E_g is the band gap energy and B is an energy-independent constant [23, 24]. The E_g is obtained by crossing the line of the function of hv. From the Tauc's plot, the band gap energy of ZnO thin films on glass substrate is approximately 3.17 eV while for ITO substrate is 3.00 eV. These results are valid since it is consistent with published values of the ZnO transition band gap data [25]. The difference of band gap energy between different substrates is influenced by the optical scattering between grain boundaries and the grains that agglomerates together in the thin films. The difference between the band gap energy of each substrate may be beneficial in many applications such as sensors, solar cells and many more.

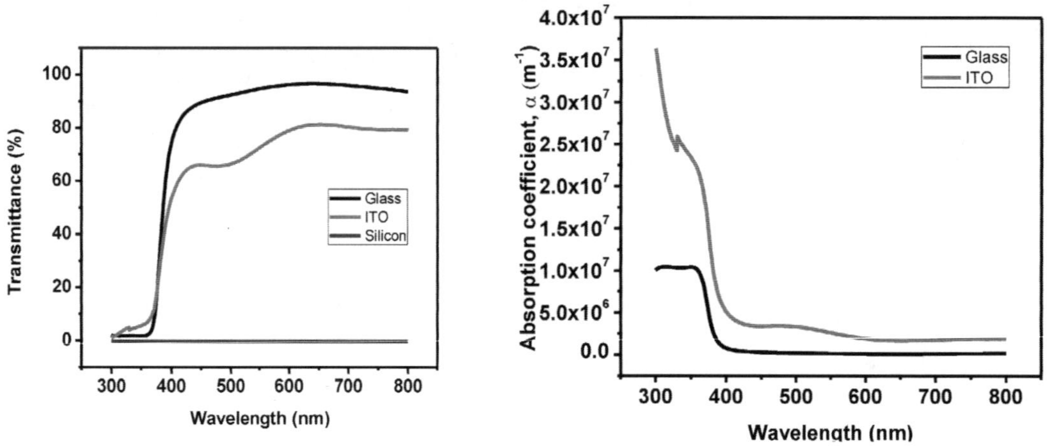

Figure 2. Transmittance spectra b) and absorption coefficient of the ZnO thin films and on different substrates

CONCLUSION

We have investigated the properties of ZnO thin films deposited using dip coating method on three different substrates which are glass, ITO and silicon. The ZnO thin films on glass were less dense as compared to ITO and silicon based on their grain boundaries and uniformity. In terms of optical properties, the ZnO thin films showed a 95% transmittance on glass in visible region (380-800 nm) which was higher than ITO. In addition, the band gap of ZnO thin films was higher on glass compared to ITO because of the larger grain boundaries. It can be concluded that structural and optical properties of the ZnO thin films were influenced by the different substrates.

ACKNOWLEDGEMENT

The authors would like to acknowledge with gratitude the NANO Electronic Centre of Faculty of Electrical Engineering UiTM for the facilities.

REFERENCES

[1] N. Singh, P. Pandey, F.Z. Haque, Effect of heat and time-period on the growth of ZnO nanorods by sol–gel technique, Optik 123 (2012) 1340–1342.
[2] C. Zhang, High-quality oriented ZnO films grown by sol–gel process assisted with ZnO seed layer, J. Phys. Chem. Solids 71 (2010) 364–369.
[3] Z. Ben Ayadi, L. El Mir, K. Djessas, S. Alaya, Effect of the annealing temperature on transparency and conductivity of ZnO:Al thin films, Thin Solid Films 517 (2009) 6305–6309.
[4] R. Hong, S. Xu, ZnO:Al films prepared by reactive mid-frequency magnetron sputtering with rotating cathode, J. Mater. Sci. Technol. 26 (10) (2010) 872–877.

[5] C. Wang, B. Hu, H. Yi, The study of structure and optoelectronic properties of ZnS and ZnO films on porous silicon substrates, Optik 123 (2012) 1040–1043.
[6] H.X. Chen, J.J. Ding, X.G. Zhao, S.Y. Ma, Microstructure and optical properties of ZnO:Al films prepared by radio frequency reactive magnetron sputtering, Physica B 405 (2010) 1339–1344.
[7] M.S. Akhtar, M.A. Khan, M.S. Jeon, O.B. Yang, Controlled synthesis of various ZnO nanostructured materials by capping agents-assisted hydrothermal method for dye-sensitized solar cells, Electrochim. Acta 53 (2008) 7869–7874.
[8] A.J. Morfa, N. Kirkwood, M. Karg, Th.B. Singh, P. Mulvaney, Effect of defects on the behavior of ZnO nanoparticle FETs, J. Phys. Chem. C 115 (2011) 8312–8315.
[9] N. Jones, B. Ray, K.T. Ranjit, A.C. Manna, Antibacterial activity of ZnO nanoparticle suspensions on a broad spectrum of microorganisms, FEMS Microbiol. Lett. 279 (2008) 71-76.
[10] N. Salah, S.S. Habib, Z.H. Khan, Adnan Memic, Ameer Azam, Esam Alarfaj, Nabeel Zahed, Salim Al-Hamedi, High-energy ball milling technique for ZnO nanoparticles as antibacterial materials, Int. J. Nanomed. 6 (2011) 863–869.
[11] K. Haga, M. Kamidaira, Y. Kashiwaba, T. Sekiguchi, H. Watanabe, ZnO thin films prepared by remote plasma-enhanced CVD method, J. Cryst. Growth 214–215 (2000) 77–78.
[12] V. Tvarozek, I. Novotny, P. Sutta, S. Flickyngerova, K. Schtereva, E. Vavrinsky, Influence of sputtering parameters on crystalline structure of ZnO thin films, Thin Solid Films 515 (2007) 8756–8760.
[13] E. Sub, S.H.S. Kang, J.S. Kang, J.H. Kim, S.Y. Lee, Effect of the variation of film thickness on the structural and optical properties of ZnO thin film deposited on sapphire substrate using PLD, Appl. Surf. Sci. 186 (2002) 474–476.
[14] K.R. Murali, Preparation of sol–gel dip-coated zinc oxide thin films, J. Phys. Chem. Solids 98 (2007) 2293–2296.
[15] G. Srinivasan, N. Gopalakrishnan, Y.S. Yu, R. Kesavamoorthy, J. Kumar, Influence of post-deposition annealing on the structural and optical properties of ZnO thin films prepared by sol–gel and spin-coating method, Superlattices Microstrut. 43 (2008) 112–119.
[16] O.A. Fouad, A.A. Ismail, Z.I. Zaki, R.M. Mohamed, Zinc oxide thin films prepared by thermal evaporation deposition and its photocatalytic activity, Appl. Catal. B: Environ. 62 (2006) 144–149.
[17] Y. Yamaguchi, M. Yamazaki, S. Yoshihara, T. Shirakashi, Photocatalytic ZnO films prepared by anodizing, J. Electroanal. Chem. 442 (1998) 1–3.
[18] Ya.I. Alivov, A.V. Chernykh, M.V. Chukichev, R.Y. Korotkov, Thin polycrystalline zinc oxide films obtained by oxidation of metallic zinc films, Thin Solid Films 473 (2005) 241–246.
[19] Saliha Elmas, Şadan Korkmaz, Deposition of Al doped ZnO thin films on the different substrates with radio frequency magnetron sputtering, Journal of Non-Crystalline Solids, Volume 359, 1 January 2013, Pages 69-72.
[20] Hu, Y., Diao, X., Wang, C., Hao, W., & Wang, T. (2004). Effects of heat treatment on properties of ITO films prepared by rf magnetron sputtering.Vacuum, 75(2), 183-188.
[21] Prepelita, P., Medianu, R., Sbarcea, B., Garoi, F., & Filipescu, M. (2010). The influence of using different substrates on the structural and optical characteristics of ZnO thin films. Applied surface science, 256(6), 1807-1811.
[22] J. Tauc, R. Grigorovici, A. Vancu, Phys. Status Solidi 15 (1996) 627.
[23] E.A. Davis, N.F. Motta, Philos. Mag. 22 (1970) 903.
[24] S.A.B.S.A. Bux, S.H.A. Aziz, Z.A. Talib, W.M.D.W. Yusoff, Solid State Sci. Technol. 13(2005) 251.
[25] J.R. Báez, A. Maldonado, G.T. Delgado, R.C. Pérez, M.D.L.L. Olvera, Mater. Lett. 60(2006) 1594.

IOP Conf. Series: Materials Science and Engineering **340** (2018) 012003 doi:10.1088/1757-899X/340/1/012003

The effects of different precursor in sonicated immersion technique on hematite nanostructure properties

W R W Ahmad[1], M H Mamat[1,2], A S Zoolfakar[1], Z Khusaimi[2], M M Yusof[1], A S Ismail[1], S A Saidi[1], M Rusop[1,2]

[1]NANO-ElecTronic Centre (NET), Faculty of Electrical Engineering, Universiti Teknologi MARA (UiTM), 40450 Shah Alam, Selangor, Malaysia

[2]NANO-SciTech Centre, Institute of Science, Universiti Teknologi MARA, 40450 Shah Alam, Selangor, Malaysia

Email: rosmaria@salam.uitm.edu.my

Abstract. In this work, hematite nanostructures were synthesized through sonicated immersion technique to study the effects of using different precursor on the surface morphologies and the optical properties of the hematite nanostructures. The surface morphologies of the samples have shown different shape of nanostructures grown on the FTO substrates that were probably affecting their optical properties. The ferric nitrate sample has shown granular shape and porous nanostructure and it has a better absorbance in the lower range of wavelength (300-550 nm), meanwhile ferric chloride has been found to cultivate a nanorod-based structure and dominant in the transmittance properties at higher wavelength (550-1200 nm).

1. Introduction

Hematite (α-Fe_2O_3) is II-VI group compound semiconductor becomes a promising material for electronic device applications due to its valuable properties. Hematite (α-Fe_2O_3) is an n-type semiconductor and the most stable iron oxide with energy band gap E_g=2.1eV in ambient condition. In aqueous solution, it has a good stability, non-toxic, corrosion resistance and abundance. α-Fe_2O_3-based material have been studied for use in many applications, including photo electrochemical water splitting[1]–[3], magnetic devices[4], [5], lithium ion batteries[4]–[7], gas sensors[8], [9], photocatalytics[10], [11], and humidity sensors[12]–[14]. There were a lot of works in controlling the growth of nanomaterials have been implemented to improve the structural and optical properties of hematite nanostructures. Some of the efforts were about to prepare the hematite nanostructures with variety of techniques include spin-coating deposition solution[15], ultrasonic spray pyrolysis[16], electrochemical deposition[17], sol-gel[18], and hydrothermal synthesis[19]. Based on previous works, a few synthesis of hematite nanostructures that were using ferric nitrate ($Fe(NO_3)_3{\cdot}9H_2O$) as precursor in solution synthesis have produced porous nanostructures[20][21] which might increase the possibility of absorbing lights or water due to related applications. However, most research works on preparing hematite nanostructures were using ferric chloride in order to produce nanorod-based structures [3], [22], [23] with regards of different stabilizer and solution-based techniques.

The main purpose of this research is to study the effects on surface morphology and optical properties of hematite nanostructures that are prepared using different precursor materials in sonicated immersion technique. In the prepared aqueous solution, we compare the use of 0.2 M ferric nitrate nonahydrate ($Fe(NO_3)_3{\cdot}9H_2O$) and 0.2 M ferric chloride ($FeCl_3{\cdot}6H_2O$) as the precursor, 0.2 M urea

(NH_2-$CONH_2$) as stabilizer, and deionized water as the solvent at identical molarity. The solutions are going through solution synthesis, which is sonicated immersion process. Then, the prepared nanostructures are characterized to investigate the difference between these two samples in terms of morphology and the optical properties.

2. Methodology

In this study, hematite nanostructures samples were divided into two, the first sample was using 0.2 M ferric nitrate nonahydrate ($Fe(NO_3)_3 \cdot 9H_2O$) as the precursor, meanwhile the second sample was using 0.2 M ferric chloride ($FeCl_3 \cdot 6H_2O$) as the precursor. Both samples have similar composition of the precursor, stabilizer and solvent, and they were also going through similar synthesis process that was sonicated immersion method. The synthesis process was started with substrate preparation, and followed by solution preparation, and material characterizations.

2.1. Substrate preparation

In this process, fluorine doped tin oxide (FTO) coated glass was used as a substrate to grow hematite nanostructures on top of its conductive surface through sonicated immersion process. The ready 2-by-2 cm FTO substrates were cleaned for the purpose of eliminating unwanted dirty particles, oxide layers and chemicals on the surface of substrate to ensure a good quality of growth nanostructures. In the cleaning step, the substrates were sonicated in methanol for ten minutes in an ultrasound bath at 50°C. Then, the substrates were rinsed with deionized (DI) water and sonicated in DI water for ten minutes at the same temperature. The substrates were rinsed with DI water and were blown for drying using a nitrogen blow-gun. The substrates were placed in the Schott bottle for further process.

2.2. Solution preparation

The solution preparation required a few steps, involving the precursors as starting material, stabilizer and solvent. The mixed solutions initially were dissolved in DI water, and went through a sonication process in a sonication waterbath for 30 minutes at 50°C as indicated in Fig 1(a). Next, the solutions were stirred on a hotplate to create homogenous solution and were charged into a Schott vial with FTO glass substrates placed horizontally in the glass vial earlier of that, and the conductive edge facing up of the vial. The mixture solutions in the Schott vial were going through sonication immersion in a water bath for 4 hours at 95°C as shown in Fig 1(b). The samples with nanostructured films formed on FTO-coated substrate were rinsed with DI water and dried at room temperature prior to a thermal annealing process for 1 hour at 500°C for surface modification.

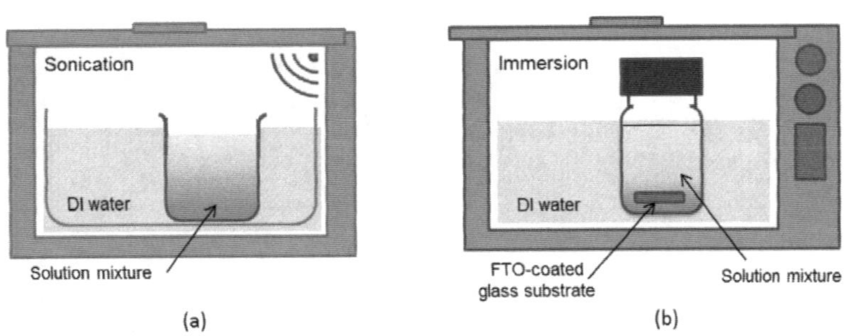

Figure 1 Experimental setup for sonicated immersion synthesis technique, (a) sonication process in an ultrasonic waterbath, and (b) immersion process in an immersion waterbath.

2.3. Material Characterizations

With the aim of analysing the surface morphology and the optical properties of the samples in different precursor, the material characterizations were conducted in two different parts; structural and optical characterizations. The field-emission scanning electron microscopy (FESEM) was used to observe the surface morphology of the hematite nanostructures, meanwhile ultraviolet-visible (UV-VIS) was used to observe the optical properties, which are absorbance and transmittance profile of the samples.

3. Results and discussions

The FESEM images in Fig 2 indicate the surface morphologies of the synthesized hematite nanostructures using two different precursors at low (30,000x) and high (50,000) magnifications. Fig 2 (a) and (b) display the hematite nanostructures that prepared using ferric nitrate as precursor that produces a granular shape and porous nanostructures. In another sample, Fig 2 (c) and (d) show the shape of nanorod hematite structure that prepared using ferric chloride as the precursor. Fig 2(c) also shows a uniform nanorod-array were grown with average diameter 67 nm and average thickness 250 nm.

The optical absorption properties within the spectrum range reveal where light is being absorbed to excite the electrons from the source of electricity. As shown in Fig 3(a), the absorbance properties of hematite nanostructures for both samples were illustrated in the range of 300-1200 nm wavelength. From the results, an intense absorbance can be seen in both samples between 300-550 nm wavelengths. However, the sample of precursor ferric nitrate has displayed higher absorbance properties compared to the sample that using precursor ferric chloride. In another optical measurement, the transmittance spectrum displays that the sample ferric chloride has a higher transparency in comparison with ferric nitrate sample as depicted in Fig 3(b). The transmittance spectrum shows a strong transparency within 550-1200 nm wavelength. The ferric chloride sample has achieved a good transmittance at visible region about 80-100% in a range of 700 to 1200 nm wavelength.

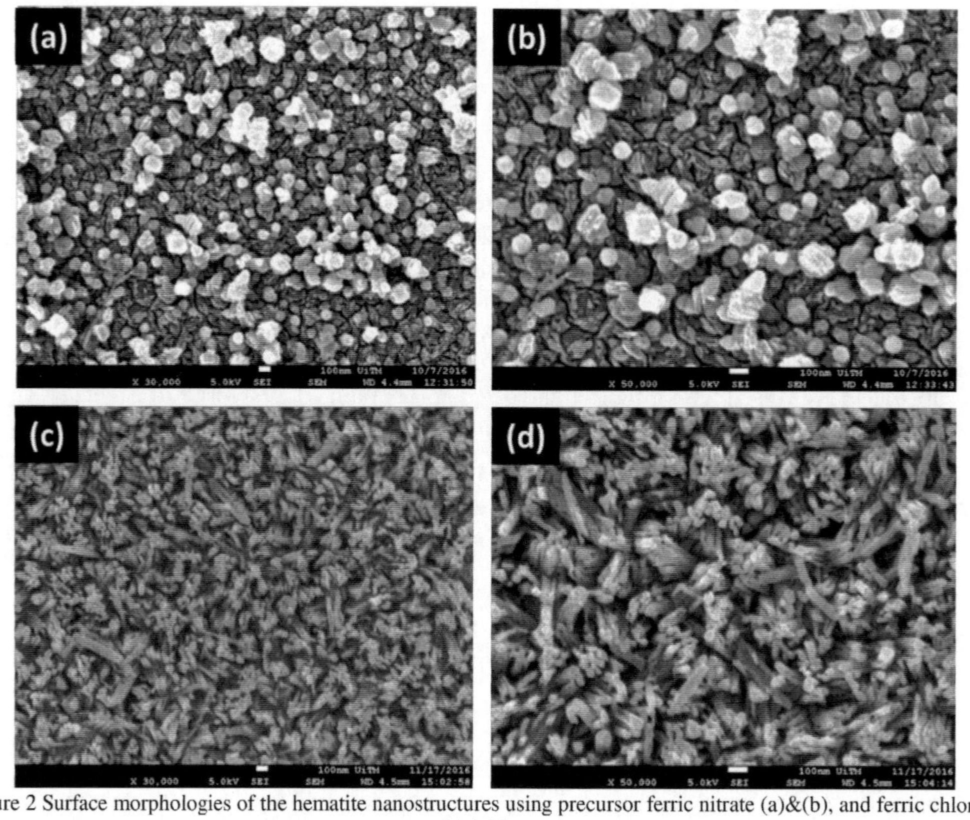

Figure 2 Surface morphologies of the hematite nanostructures using precursor ferric nitrate (a)&(b), and ferric chloride (c)&(d).

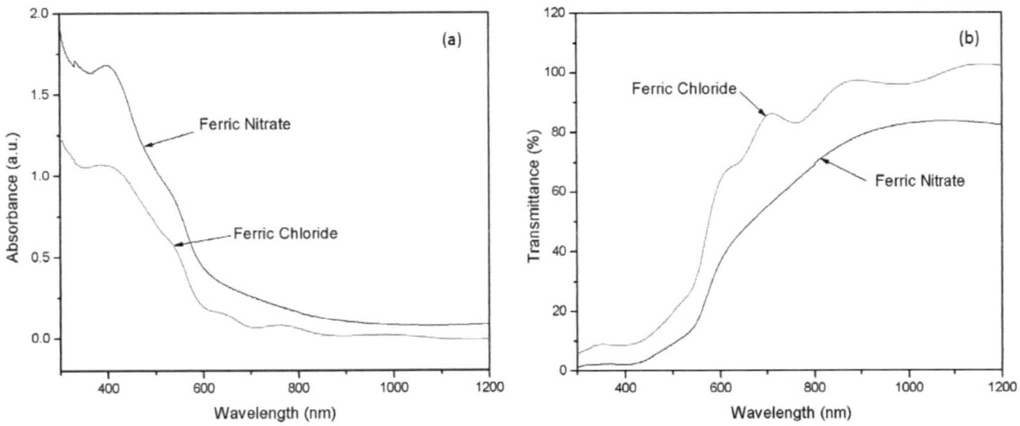

Figure 3 The optical properties of the hematite nanostructures; (a) absorbance, and (b) transmittance.

4. Conclusions

In this study, we have successfully prepared hematite nanostructures to investigate the effects of using different materials as precursor through sonicated immersion technique on the surface morphology and its optical behaviours. From the obtained results, it was found that the different precursor materials used in the immersion method would produce different shape of nanostructures on the FTO substrate. Additionally, these nanostructures also result in different behaviour of the optical properties, as we can see that the hematite nanostructure prepared using ferric nitrate has a good absorbance, meanwhile the hematite nanostructure prepared using ferric chloride offers a better transparency. In this observation, we believe that ferric nitrate causes a higher porosity of hematite granular nanostructure grown on FTO substrate, thus increases the rate of light adsorption on top of it. The nanorod profile of the nanostructures prepared using precursor ferric chloride provided a good transmittance at wavelength above 600 nm probably due to its one dimension architecture.

Acknowledgments

This research work was financially supported by the LESTARI grant (600-IRMI/MYRA5/3/LESTARI (0162/2016)) and Faculty of Electrical Engineering from Universiti Teknologi MARA, Malaysia. The authors also would like to thank the Research Management Institute (RMI) of UiTM for supporting this research work.

References

[1] R. Rajendran, Z. Yaakob, M. Pudukudy, M. S. A. Rahaman, and K. Sopian, "Photoelectrochemical water splitting performance of vertically aligned hematite nanoflakes deposited on FTO by a hydrothermal method," *J. Alloys Compd.*, vol. 608, pp. 207–212, 2014.

[2] S. Shen, J. Zhou, C.-L. Dong, Y. Hu, E. N. Tseng, P. Guo, L. Guo, and S. S. Mao, "Surface engineered doping of hematite nanorod arrays for improved photoelectrochemical water splitting.," *Sci. Rep.*, vol. 4, p. 6627, 2014.

[3] V. A. N. De Carvalho, R. A. D. S. Luz, B. H. Lima, F. N. Crespilho, E. R. Leite, and F. L. Souza, "Highly oriented hematite nanorods arrays for photoelectrochemical water splitting," *J. Power Sources*, vol. 205, pp. 525–529, 2012.

[4] N. Pariona, K. I. Camacho-Aguilar, R. Ramos-González, A. I. Martinez, M. Herrera-Trejo, and E. Baggio-Saitovitch, "Magnetic and structural properties of ferrihydrite/hematite nanocomposites," *J. Magn. Magn. Mater.*, vol. 406, pp. 221–227, 2016.

[5] M. Tadic, M. Panjan, V. Damnjanovic, and I. Milosevic, "Magnetic properties of hematite (a-

Fe2O3) nanoparticles prepared by hydrothermal synthesis method.," *Appl. Surf. Sci.*, vol. 320, pp. 183–187, 2014.

[6] M. Balogun, Z. Wu, Y. Luo, W. Qiu, X. Fan, B. Long, M. Huang, P. Liu, and Y. Tong, "High power density nitridated hematite (α-Fe2O3) nanorods as anode for high-performance flexible lithium ion batteries," *J. Power Sources*, vol. 308, pp. 7–17, Mar. 2016.

[7] L. Chen, H. Xu, L. Li, F. Wu, J. Yang, and Y. Qian, "A comparative study of lithium-storage performances of hematite: Nanotubes vs. nanorods," *J. Power Sources*, vol. 245, pp. 429–435, 2014.

[8] B. Zhang, W. Fu, X. Meng, A. Ruan, P. Su, and H. Yang, "Synthesis and enhanced gas sensing properties of flower-like ZnO/α-Fe2O3 core-shell nanorods," *Ceram. Int.*, 2017.

[9] H. J. Song, Y. Sun, and X. H. Jia, "Hydrothermal synthesis, growth mechanism and gas sensing properties of Zn-doped α-Fe2O3 microcubes," *Ceram. Int.*, vol. 41, no. 10, pp. 13224–13231, 2015.

[10] T. S. Atabaev, "Facile hydrothermal synthesis of flower-like hematite microstructure with high photocatalytic properties," *J. Adv. Ceram.*, vol. 4, no. 1, pp. 61–64, 2015.

[11] Z. Zhang, M. F. Hossain, and T. Takahashi, "Self-assembled hematite (Fe2O3) nanotube arrays for photoelectrocatalytic degradation of azo dye under simulated solar light irradiation," *Appl. Catal. B Environ.*, vol. 95, no. 3–4, pp. 423–429, 2010.

[12] A. S. Afify, M. Ataalla, A. Hussain, M. Hassan, A. Mohammed, M. Milanova, and J. M. Tulliani, "Studying the effect of doping metal ions onto a crystalline hematite-based humidity sensor for environmental control," *Bulg. Chem. Commun.*, vol. 48, no. 2, pp. 297–302, 2016.

[13] P. V Adhyapak, U. P. Mulik, D. P. Amalnerkar, and I. S. Mulla, "Low Temperature Synthesis of Needle-like alpha-FeOOH and Their Conversion into alpha-Fe2O3 Nanorods for Humidity Sensing Application," *J. Am. Ceram. Soc.*, vol. 96, no. 3, pp. 731–735, 2013.

[14] J. M. Tulliani, C. Baroni, L. Zavattaro, and C. Grignani, "Strontium-doped hematite as a possible humidity sensing material for soil water content determination.," *Sensors (Basel).*, vol. 13, no. 9, pp. 12070–12092, 2013.

[15] F. L. Souza, K. P. Lopes, P. A. P. Nascente, and E. R. Leite, "Nanostructured hematite thin films produced by spin-coating deposition solution: Application in water splitting," *Sol. Energy Mater. Sol. Cells*, vol. 93, no. 3, pp. 362–368, 2009.

[16] S. S. Shinde, R. a Bansode, C. H. Bhosale, and K. Y. Rajpure, "Physical properties of hematite α-Fe 2 O 3 thin films: application to photoelectrochemical solar cells," *J. Semicond.*, vol. 32, no. 1, p. 13001, 2011.

[17] Q. Meng, Z. Wang, X. Chai, Z. Weng, R. Ding, and L. Dong, "Fabrication of hematite (α-Fe2O3) nanoparticles using electrochemical deposition," *Appl. Surf. Sci.*, vol. 368, pp. 303–308, 2016.

[18] U. Khan, A. Akbar, H. Yousaf, S. Riaz, and S. Naseem, "Ferromagnetic Properties of Al-doped Fe2O3 Thin Films by Sol-gel," *Mater. Today Proc.*, vol. 2, no. 10, pp. 5415–5420, 2015.

[19] A. Pu, J. Deng, Y. Hao, X. Sun, and J. Zhong, "Thickness effect of hematite nanostructures prepared by hydrothermal method for solar water splitting," *Appl. Surf. Sci.*, vol. 320, pp. 213–217, 2014.

[20] X. Gou, G. Wang, J. Park, H. Liu, and J. Yang, "Monodisperse hematite porous nanospheres: synthesis, characterization, and applications for gas sensors.," *Nanotechnology*, vol. 19, no. 12, p. 125606, 2008.

[21] I. Cesar, A. Kay, J. A. G. Martinez, and M. Grätzel, "Translucent thin film Fe2O3 photoanodes for efficient water splitting by sunlight: Nanostructure-directing effect of Si-doping," *J. Am. Chem. Soc.*, vol. 128, no. 14, pp. 4582–4583, 2006.

[22] H. K. Mulmudi, N. Mathews, X. C. Dou, L. F. Xi, S. S. Pramana, Y. M. Lam, and S. G. Mhaisalkar, "Controlled growth of hematite (Fe2O3) nanorod array on fluorine doped tin oxide: Synthesis and photoelectrochemical properties," *Electrochem. commun.*, vol. 13, no. 9, pp. 951–954, 2011.

[23] Gurudayal, S. Y. Chiam, M. H. Kumar, P. S. Bassi, H. L. Seng, J. Barber, and L. H. Wong, "Improving the efficiency of hematite nanorods for photoelectrochemical water splitting by doping with manganese.," *ACS Appl. Mater. Interfaces*, vol. 6, no. 8, pp. 5852–9, Apr. 2014.

In-situ Growth of Sn-Doped ZnO Nanotetrapods using Thermal CVD Method

S.S. Shariffudin, N.E.A. Azhar, A.Z. Rodzuan, M.S.P. Sarah, H. Hashim

Faculty of Electrical Engineering, Universiti Teknologi MARA, 40450 Shah Alam, Selangor

sobihana@salam.uitm.edu.my

Abstract. Tin (Sn) insitu doping of Zinc Oxide (ZnO) nanotetrapods (T-SZnO) were successfully grown using thermal chemical vapour deposition (TCVD) method. The objective of this paper is to study the surface morphology, optical and electrical properties of the samples by varying the atomic percentage of the Sn during the deposition. All samples were annealed at temperature of 500°C for 1 hour. Field emission scanning electron microscopy (FESEM) images showed that diameter and length of nanotetrapods increased with the dopant concentration. Energy dispersive x-ray spectroscopy (EDX) showed that the nanotetrapod ZnO had been doped with Sn. Red shift was observed in the visible blue emission and a decrease in both UV and visible peaks for the photoluminescence spectra. Current-voltage (*I-V*) measurement showed the current increased with the dopant concentration.

1. Introduction

Wide band gap metal oxide semiconductors such as ZnO have been extensively studied since they showed great properties for many electronic devices. ZnO is a II-VI semiconductor compound with a direct band gap of E_g = 3.37 eV (at room temperature) [1, 2] which makes it suitable for application of shortwave length optoelectronics devices, such as light emitting diode and laser diodes. ZnO has high excitation binding energy of 60 meV, which is three times larger than GaN (E_b = 20 meV). Besides that, ZnO has good transparency, high electron mobility and also strong room temperature luminescence. The properties of this material depend on the deposition techniques, deposition parameter, post deposition treatment of the films and doping. The nanostructure of this material can be deposited by variety of techniques such as radio frequency (RF) sputtering, sol- gel method, thermal oxidation of metallic zinc, thermal evaporation, pulsed laser deposition (PLD) and spray pyrolysis. The nanostructure of ZnO can be found in various morphologies including nanowires, nanorods, nanotetrapods, nanobelts, nanoflowers, nanopartical and many more. Among the various nanostructures, nanotetrapod has a unique structure which shows improved characteristics compared to others. Many devices had proven better performance when incorporated ZnO nanotetrapods, such as in gas sensors [3] and UV sensors. It also improved the polymer performance when combined into the nanocomposite thin films as fillers.

Doping is an important and effective way to improve the properties of metal-oxide semiconductors. Even though many researchers have reported on the doping of chemical vapour deposited ZnO using metal ions such as Sn [4],Ti [5], Sb [6], Mn [7] and many more which mostly focuses on the nanorods but only a few of them produced nanotetrapods.

The aim of this study is to investigate on the growth of ZnO nanotetrapods with Sn in-situ doping using thermal CVD method. The influence of different dopant concentration on surface morphology, optical properties and electrical properties of the nanostructure were characterized by FE-SEM, PL and *I-V*.

2. Experimental

2.1. Samples preparation
Glass substrates were cut with a measurement of 2.5cm x 2.5cm dimension. The glass substrates was cleaned using acetone, methanol and deionized (DI) water in ultrasonic for 10 minutes each. Then, the glass substrates were dried with nitrogen gas.

The nanotetrapod T-SZnO was synthesized using a double furnace TCVD. 2g of Zn powder ($\geq$99.9%, Sigma Aldrich) and different doping concentration of Sn was placed in two alumina boats at furnace 1 and another empty alumina boat was placed in furnace 2 for collecting the deposited sample. These boats were inserted into the quartz tube within certain diameter. Before the growth, quartz tube was put into the center area of horizontal furnace tube with the facing the direction of gas flow. Argon (Ar) which acts as the carrier gas was set at 70 sccm, while oxygen (O_2) gas as the reactor gas and was set at 5 sccm. The deposition temperature was set at 750 °C for furnace 1 while furnace 2 was set at 500 °C. It took 30 minutes to complete the deposition process. The white cottony product was collected once the furnace was cooled to room temperature.

The method for preparation of thin films for characterization is described as follows: A few drops of methanol solution were put into the beaker filled with the T-SZnO and stirred. A drop of solution on the substrate was rolled by a roller until the solution become uniform surface. The thin film was dried at 150°C for 10 minutes. Lastly, the nanotetrapods T-SZnO thin films were annealed at 500°C for 1 hour.

2.2. Characterization
Surface morphology and material composition of nanotetrapods T-SZnO thin film was characterized using FESEM and EDX (JEOL JSM-7600F). The optical properties were measured using photoluminescence emission (PL). The electrical properties T-SZnO thin films were measured using two-point probe solar simulator system (CEP 2000).

3. Result and Discussion

3.1 Surface Morphology
The surface morphology of the T-SZnO and undoped ZnO nanotetrapod was observed using field-emission scanning electron microscope (FE-SEM). The FE-SEM image was taken at 50 k magnification for 5kV, as shown in Figure 1. Figure 1 (a), (b), (c), and (d) shows the different sizes of nanotetrapods T-SZnO thin film due to different of Sn dopant concentrations. Meanwhile, Figure 1 (e) shows the size of the undoped ZnO nanotetrapods sample. It can be seen that the surface of thin films exhibit the nanotetrapod shape, where each one consists of four hexagonal rod-shaped legs connected at the center tetrahedrally. The hexagonal terminal of each leg indicates the wurtzite phase of T-SZnO crystal.

The diameter size of thin films was summarized in Table 1. The average size of the T-SZnO was found to increase with the Sn dopant concentration. The length was between 380 nm to 588 nm due to different Sn dopant concentration as shown in Table 1 as compared to the sample of undoped ZnO the length of the nanotetrapod was 330nm. Based on the result, the sizes of nanotetrapod T-SZnO was much bigger than the size of undoped ZnO. This proves that the changes in morphology of the T-SZnO nanotetrapods might due to the substantive influence of Sn-ion into the ZnO crystals.

IOP Conf. Series: Materials Science and Engineering **340** (2018) 012004 doi:10.1088/1757-899X/340/1/012004

Figure 1. FESEM images of nanotetrapod T-SZnO films at (a) 2%, (b) 4%, (c) 6%, (d) 8%, (e) undoped

Table 1. The length and diameter of T-SZnO samples at different Sn concentrations

Sn Concentration (at%)	Length (nm)	Diameter (nm)
0	330	72.9
2	380	110
4	400	115
6	551	121
8	588	126

The chemical compositions of the T-SZnO thin film were measured using energy dispersive x-ray spectroscopy (EDX). The EDX spectrum of nanotetrapods T-SZnO thin film is shown in Figure 2 and it is confirmed that the nanotetrapods contains Sn which showed that the T-SZnO was successfully doped during the growth process. The EDX spectrum shows that the major compositions for T-SZnO nanotetrapod thin films are Sn (0.48 %), Zn (44.40 %) and O (50.00 %).

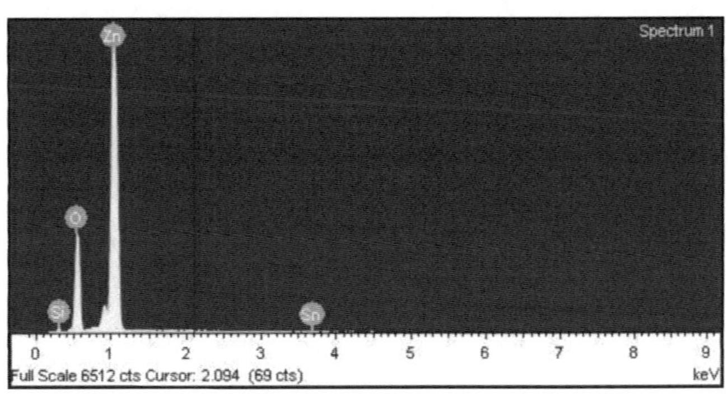

Figure 2. EDX spectra of nanotetrapod T-SZnO

3.2 Optical Properties

The optical properties of nanotetrapod T-SZnO thin film was observed by photoluminescence (PL) emission at room temperature. The wavelength was taken in the range of 350 to 700 nm. Figure 3 shows the PL spectra of T-SZnO at different concentration of Sn. Based on the PL results, it is noticed that there are peaks at ultra violet (UV) emission at about 390 nm wavelength with a secondary blue peak next to it. The UV band is corresponding to the near band-edge (NBE) emission, while for visible emission is normally related to ionized oxygen vacancies and defect. Interestingly the intensity of both UV and blue emission decrease with the increase of the Sn doping. Besides that, it is clearly observed that both UV and blue emission shifted to higher wavelength side (from 406 nm for sample 2% to 420 nm for sample 8%) in addition to the decrease of intensity. This red shift of the UV emission indicates a reduction of the ZnO band gap caused by the Sn doping. The red shifting of PL was also reported by Ye et. al. [8]. According to Ameen et. al [9], the Sn-ion might take part in the substitution of Zn ion which resulted in red shifting of the visible emission.

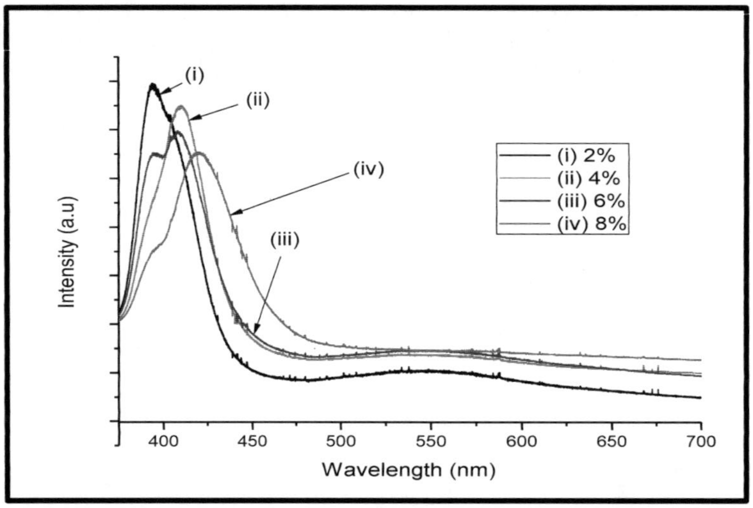

Figure 3. PL spectra of nanotetrapod T-SZnO thin film at i) 2%, (ii) 4%, (iii) 6%, and (iv) 8%

3.3 Electrical Properties

The current - voltage (I-V) measurement was measured to study the effect on the electrical properties of T-SZnO thin film. The *I-V* measurement of nanotetrapod T-SZnO thin film was measured using two-point probe solar simulator within voltage of -8 V to 8 V. Figure 4 shows the current-voltage of nanotetrapod T-SZnO thin films with different Sn concentrations at room temperature. The (*I-V*) curves are rectifying which indicates that the gold metal contact was reacting with the thin films.

The graph indicates that the current of T-SZnO increased with the concentration of Sn doping, which showed that the dopant improves the electrical characteristic of the thin films. We can state that the increases of dopant concentration might increase the current intensity because of increase of electron concentration and electron mobility [8]. Besides that, it might have contributed by the reduction of the thin films' band gap.

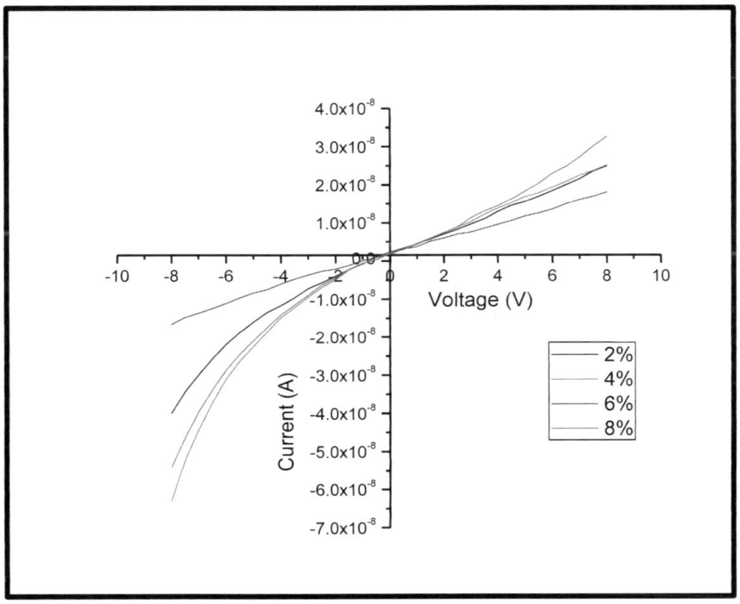

Figure 4. *I-V* measurements of nanotetrapod T-SZnO thin film different Sn concentration

4. Conclusion

A high quality Sn doped ZnO (T-SZnO) nanotetrapods were successfully synthesized using thermal-CVD method with different Sn concentration. The effect of Sn doping in ZnO indicates significant morphological changes whereby the FESEM results showed that there is an increased in the nanotetrapods sizes. This reveals the incorporation of Sn-ion into ZnO crystals. Similar result is also shown by the red shifting of the visible emission of the PL spectra. The *I-V* curve showed the current of thin film increased with the increase of the dopant concentration.

Acknowledgement

We would like to express our gratitude to Ministry of Higher Education Malaysia under FRGS grant (600-RMI/FRGS 5/3 (119/2015), and Institute of Research Management and Innovation (IRMI), Universiti Teknologi MARA (UiTM), Shah Alam for financial support. We would like to express our gratitude to NANO-ElecTronic Centre (NET), Faculty of Electrical Engineering, and NANO-SciTech Centre, Universiti Teknologi MARA, Shah Alam for making the data available for study.

References

[1] A. B. Djurišić, A. M. C. Ng, and X. Y. Chen, *Progress in Quantum Electronics,* vol. 34, pp. 191-259, 2010.
[2] I.-M. Chan and F. C. Hong, *Thin Solid Films,* vol. 450, pp. 304-311, 2004.
[3] M. Carotta, A. d. Cervi, V. Di Natale, S. Gherardi, A. Giberti, V. Guidi, *et al Sensors and Actuators B: Chemical,* vol. 137, pp. 164-169, 2009.
[4] X. Zhou, T. Lin, Y. Liu, C. Wu, X. Zeng, D. Jiang, *et al., ACS applied materials & interfaces,* vol. 5, pp. 10067-10073, 2013.
[5] K. Zheng, L. Gu, D. Sun, X. Mo, and G. Chen *Materials Science and Engineering: B,* vol. 166, pp. 104-107, 2010.
[6] Q. J. Feng, S. Liu, Y. Liu, H. F. Zhao, J. Y. Lu, K. Tang, *et al., Materials Science in Semiconductor Processing,* vol. 29, pp. 188-192, 2015.
[7] H. Yan, X. Zhong, J. Wang, G. Huang, S. Ding, G. Zhou, *et al., Applied physics letters,* vol. 90, p. 082503, 2007.

[8] N. Ye, J. Qi, Z. Qi, X. Zhang, Y. Yang, J. Liu, *et al.*, *Journal of Power Sources,* vol. 195, pp. 5806-5809, 2010.
[9] S. Ameen, M. S. Akhtar, H.-K. Seo, Y. S. Kim, and H. S. Shin, *Chemical Engineering Journal,* vol. 187, pp. 351-356, 2012.

IOP Conf. Series: Materials Science and Engineering **340** (2018) 012005 doi:10.1088/1757-899X/340/1/012005

Fabrication of p-type Organic Field Effect Transistor using PMMA:TiO_2 as Nanocomposite Dielectric Layer

L. N. Ismail*, Samsul S., Musa M. Z, S. Norsabrina

Faculty of Electrical Engineering, Universiti Teknologi MARA Cawangan Pulau Pinang, 13500 Seberang Perai, Pulau Pinang

E-mail : *lyly2909@gmail.com

Abstract. This paper report on the fabrication of p-type organic field effects transistor (OFET) using low temperature spin coating deposition technique. The p-type OFETs were fabricated using poly (3-hexylthiophene-2,5-diyl) (P3HT) and poly (methyl methacrylate):titanium dioxide (PMMA:TiO_2) nanocomposite material were used as semiconductor and the gate dielectric, respectively. The electrical and structural properties of OFETs were analysed using current-voltage (I-V), atomic force microscopy (AFM) and RAMAN spectroscopy. Results from I-V show that the fabricated p-type OFET has moderate performance with threshold voltage, V_{TH} is -3V and mobility, μ is 2.01 $cm^2 V^{-1}s$.

1. Introduction

Organic field effect transistor (OFET) is an example of organic electronics devices and has attracted a great deal of attention to be utilized in integrated circuit (IC) [1]. OFET structure consists of semiconductors, dielectric layer and electrodes that fabricate using organic or polymer material. OFETs is suitable to be used for low voltage application such as smart labels, driving matrices for organic flat panel displays [2], sensor [3] and food industry [4]. The innovation of OFET has grown in recent years and have been focused on the development of improving the properties of organic semiconductor materials and methods for improving the material to achieve high carrier mobility [5, 9-11]. Carries mobilities is an important parameter in determining and enhancing the OFET performance because OFET basically operates through the modulation of charge carrier flow at the interface of semiconductor-dielectric.

In order to investigate the OFET performance, metal-insulator-semiconductor (MIS) structure was used to investigate the behaviour between the insulator and semiconductor. The performance of MIS capacitor is strongly depend on semiconductor and dielectric layer. The important characteristics for MIS is the accumulation/depletion behaviour, charge transport and trapping, interface effects between insulator and semiconductor [1]. In order to obtain good MIS characteristics, the interface between insulator and the semiconductor layer must be well-matched because the insulator were sandwiched between the semiconductor and ohmic metal contacts and the bias voltage is being applied across an insulator. The same characteristics and principles was applied to the MIS that fabricated using organic material in the case poly (3-alkylthiophene) (P3HT) and poly (methyl methacrylate): titanium dioxide (PMMA:TiO_2).

(P3HT) [18] is a stable p-type organic semiconductor material and the most prominent organic semiconductor material that has been widely being used by other researcher in fabricating the organic electronic devices. Polythiophenes, in particular (P3HT), are the most prominent and investigated representatives of semiconducting polymers and have been applied in various devices such as solar cells and field-effect transistors. For this class of polymers, it has been well established that the morphology of the functional layer has a significant impact on the device performance. For insulator layer, the nanocomposite material between PMMA and TiO_2 was used due to its high dielectric permittivity, ε' that has been optimized from our previous work. Besides having high dielectric permittivity, (ε' ~ 12), PMMA:TiO_2 also have high capacitance for storage purpose and it can be processor at low temperature [12].

In this paper, we report the fabrication and characterization of metal-insulator-semiconductor (MIS) and p-type organic field effect transistor (p-OFET) using PMMA:TiO_2 nanocomposite and P3HT as insulator and semiconductor layer, respectively. Both thin film was deposited by spin coating onto glass substrate and characterized by current-voltage (I-V), atomic force microscopy (AFM) and RAMAN spectroscopy. In this work, we only concentrate on the I-V characteristics of MIS and p-type OFET.

2. Experimental Setup

Before coating, the glass substrate was clean by sonication method in acetone, methanol and de-ionized (DI) water in ultrasonic bath for 10 min at 50 °C, then dried using nitrogen (N_2) gas.

2.1. Fabrication of PMMA:TiO_2 Nanocomposite dielectric film

PMMA:TiO_2 solution was prepared by dissolving 0.3g of PMMA (Sigma Aldrich) and 3wt% of TiO_2 nanopowder in 5ml of Toluene. A few drops of TMOMS were added into PMMA: TiO_2 as stabilizer between PMMA and TiO_2 nanopowder. Then, the solution was sonicated for 30 min at 50 °C before being stirred for 24 hrs. The deposition of PMMA: TiO_2 thin film was done in a spin coater with argon (Ar) gas at 4 mbar, 0.5 s/lit. The time and spin speed of spin coater were fixed to 4000 rpm for 60 sec in order to obtain the uniform PMMA: TiO_2 nanocomposite film. 2 drops of PMMA: TiO_2 solution was dropped onto the substrate and the rotation would evenly spread the solution, and stop automatically after 60 sec. The substrate with a layer of 2 drops was then dried at 50 °C for 5 minutes to evaporate the solvent and harden the coating. After that, the PMMA: TiO_2 films were annealed at 120 °C for 30 minutes, enough to sufficiently remove the remaining solvent.

2.2. Fabrication of P3HT Thin Film

P3HT solution was prepared by dissolving 15mg of P3HT powder (Sigma Aldrich) in 15 ml of chloroform solvent (Eriendemann Schmidt). The solution was sonicated for 30 min at 50 °C before being stirred for 24 hrs. The P3HT film was deposited using spin coating technique. The spin speed was fixed at 1500 rpm for 60 sec. Then, the P3HT film is dried at 50 °C for 5 min to evaporate the solvent.

2.3. Fabrication of P3HT Thin Film

The MIS (Al/P3HT/PMMA:TiO_2/Al) structure as shown in Fig. 1 were fabricated to investigate the performance of OFETs. The MIS device were fabricate on top of glass substrate. 60 nm thick of aluminum (Al) were evaporate on top of glass substrate to form bottom electrode. P3HT solution were spin coated on top of Al electrode using the same technique to form P3HT thin film. Then, PMMA:TiO_2 nanocomposite dielectric film were deposited on top of P3HT thin film followed by deposition of 100 nm thick Al to form top electrode.

The fabrication of p-type OFET start by depositing the P3HT semiconductor layer on glass substrate. Al source and drain terminals were deposited using thermal evaporator technique. Then, a 500 nm thick of PMMA:TiO_2 nanocomposite dielectric were deposited on top of source and drain terminal act as dielectric layer. Then, Al gate electrode with 100 nm thick were deposited on top of PMMA:TiO_2 layer. Complete p-type OFET structure as shown in Fig. 2 below.

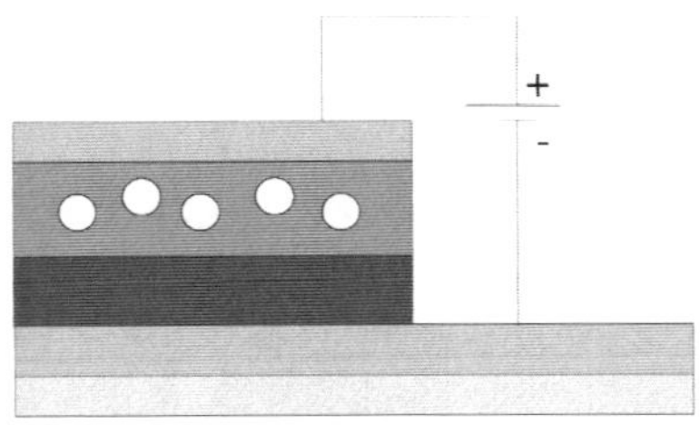

Figure 1. Schematic structure of the MIS

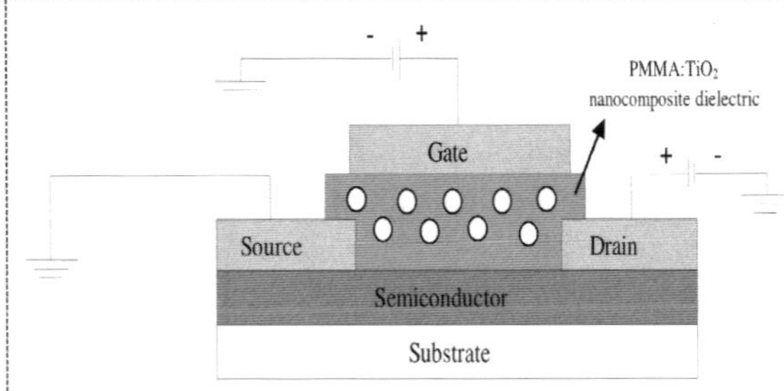

Figure 2. Schematic diagram of the OFETs

3. Results and Discussions

3.1. Structural Properties

The Raman spectrum of PMMA:TiO_2 nanocomposites and P3HT thin film are shown in Fig. 3. It can be seen that there is most of the Raman bands for PMMA:TiO_2 are identified are the same as Raman band for PMMA that has been reported in the literature [21,22]. The spectrums for PMMA:TiO_2 occur at 2957, 1460 cm^{-1} and weak features at 3001 cm^{-1} are belonging to the C-H vibration. The features at 2848 cm^{-1} is corresponding to the infrared spectrum [23]. The feature at 999 cm^{-1} was assigned to OCH3 rock, the feature at 1081 cm^{-1} to v(C-C) skeletal mode vibration. The observed bands are at 602, 1460 and 1736 cm-^{1}. A new peaks which can be seen in PMMA:TiO_2 curve are at 141 cm^{-1} indicate the presence of TiO_2 nanoparticles which associated with rutile phase [24]. The 715 cm^{-1} , 1380 cm^{-1} and 1440 cm^{-1} Raman peaks indicated by squares are assigned to the P3HT C–S– C ring deformation, C–C skeletal stretching and CLC ring stretching, respectively.

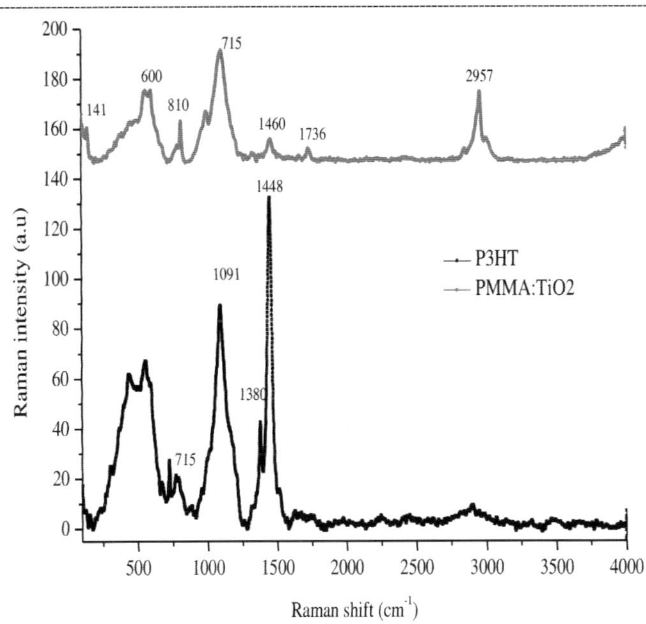

Figure 3. RAMAN spectrum of P3HT and PMMA:TiO_2 film.

Figure 4 and Fig.5 is an FESEM image displaying the surface morphology of PMMA:TiO_2 nanocomposite films and cross-section image of P3HT/PMMA:TiO_2 thin film on glass substrate. As can be observed, the surface of the nanocomposite PMMA:TiO_2 is not smooth and there is agglomerated

IOP Conf. Series: Materials Science and Engineering **340** (2018) 012005 doi:10.1088/1757-899X/340/1/012005

particles at certain area. This agglomerated particle is confirmed that there is a formation of non-homogenous organic-inorganic materials with very small domain in the nanometric scales. The root mean square (rms) roughness of the film measured in this area is 72.903 nm for nanocomposite PMMA:TiO_2. The topography of P3HT films were observed by AFM as shown in Figure 6. The surface roughness in root mean square (RMS) data is 0.94 nm. This suggests that the P3HT molecules are slightly more densely packed, and hence slow down the diffusion of oxygen and moisture to the gate dielectric interface where the holes transport takes place. This may also contribute to the improved stability for the P3HT devices based on PMMA:TiO_2 coated substrates.

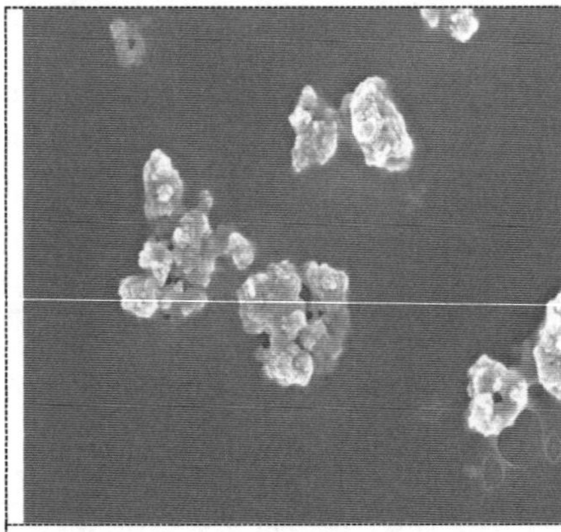

Figure 4. FESEM images of PMMA:TiO_2 thin film.

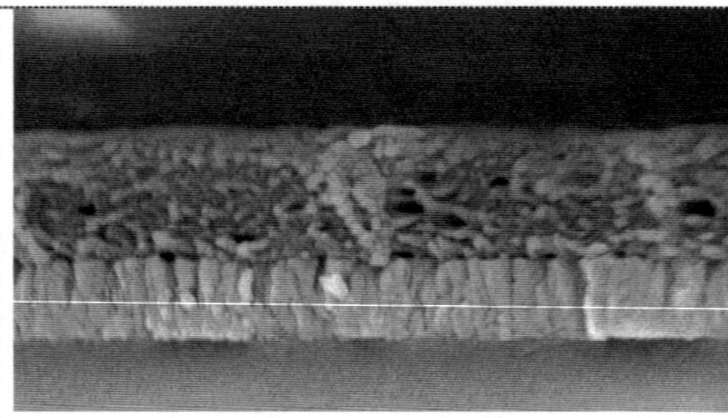

Figure 5. Cross-section image of P3HT and PMMA:TiO_2 film.

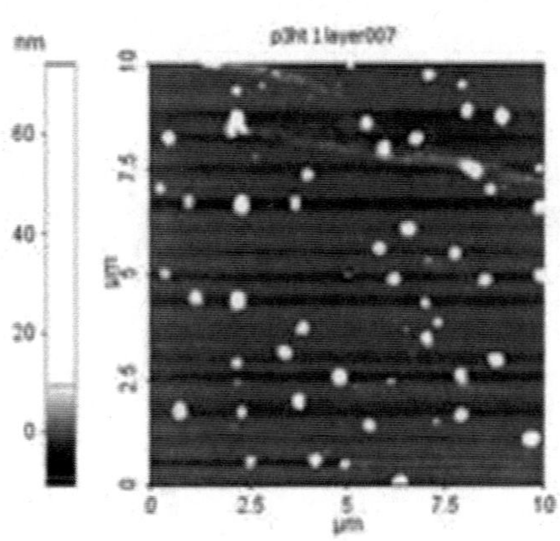

Figure 6. AFM image of P3HT thin film.

3.2. Electrical Properties of MIS and p-type OFET

The purpose of fabricating the MIS devices is to investigate the suitability and the compatibility of fabricated PMMA:TiO_2 nanocomposite dielectric layer with p-types semiconductor which is P3HT film. I-V characteristics of MIS with different annealing temperature of P3HT film are shown in Fig.7 below. The purpose of varying the annealing temperature is to obtain the optimization temperature to deposit the P3HT thin film for OFET fabrication. The applied voltages were sweep from +10 V to -10 V to the Al bottom electrode while top Al electrode was grounded. I-V curve for all MIS samples show nonlinear or rectifying behaviour with different operating voltages. The operating voltages are increased significantly when the annealing temperature of P3HT film increased. MIS device with 60 °C has the lowest operating voltage ~ - 7 V and MIS with 180 °C has the highest operating voltage which is – 1.0

V. Increment in the operating voltage, when different annealing temperature was used is because of the changes in the surface roughness and also reduction of P3HT thin film. Since the surface roughness is reduce, when the negative voltage is applied (V > -1 V) to the Al electrode, the majority carriers (hole) are accumulated near the surface of the P3HT layer. This causes the valence band edge bends downwards near the surfaces and closer to the Fermi level. When a small positive voltage (V < 0 V) is applied, the majority carriers (hole) are depleted because the valence band bends upward. The bands bend even more upward when applied large positive voltage (V < +2 V) so that the intrinsic Fermi level crosses over the Fermi level and at this point the minority carrier (electron) is larger than hole, the surface thus inverted. These excess electrons caused increased current in the positive.

The electrical properties of p-type OFET were characterize using 4200 semiconductor characterization system. Figure 8 shows the typical drain-source current, I_{DS} characteristics of p-type OFETs with PMMA:TiO_2 and P3HT as dielectric and semiconductor layer. The drain-source voltage, V_{DS} was varied from 0 ~ - 4 V in steps 0.1. For this p-type OFETs, negative V_{DS} were applied because P3HT is a p-type semiconductor/active material, therefore the majority carriers are holes. Figure 8 shows that the applied gate voltage, V_G was varied from -10 to +10 V. The I_{DS}-V_{DS} curves indicate that our fabricated p–type OFET, resembled the theoretically transistor curves, nevertheless, the curves are not ideal. This is due to the parallel resistance that exists in the device. It can be also seen that, apparently that the fabricated p-type OFET is in the depletion mode (normally-ON) state. When the no gate voltage, V_G (V_G = 0V) is applied, the fabricated p-type OFETs are in ON state. For these devices, since the majority carries is holes, therefore the holes channel is induced at the semiconductor-dielectric (which is P3HT- PMMA:TiO_2) interface by a negative bias applied to the gate. The current flows form source to drain electrodes via holes channel if the $-V_G$ is sufficiently high. At this condition, the channel acts as variable resistors and the I_{DS} are linear with $-V_{DS}$, which correspond to the linear region. As the $-V_{DS}$ increases, the space charge region under the channel and near the drain region widens and eventually reaches a pinch-off point condition in which the channel width becomes zero. Beyond this pinch-off point, in the saturation region, the I_{DS} become saturated or constant as $-V_{DS}$ continues to increase. By applying positive V_G, the p–type OFET start to change state from ON to OFF state. The fabricated p-type OFET in particular, there is no good saturation current observed. Figure 9 illustrated the I_{DS} versus V_{GS} by fixing the V_{DS} at -10 V. From these curves, the field effect mobility, μ_{FE} and the threshold voltage, V_{TH} were determined. The μ_{FE} was extracted from the slope of the linear plots of the square root of the I_{DS} versus the V_G. The V_{TH} was estimated corresponds to the intersection dashed lines with V_{GS} axis is - 1V. Meanwhile, the μ_{FE} of the p-type OFETs is 8 x 10^{-3} cm2 /Vs^{-1}.

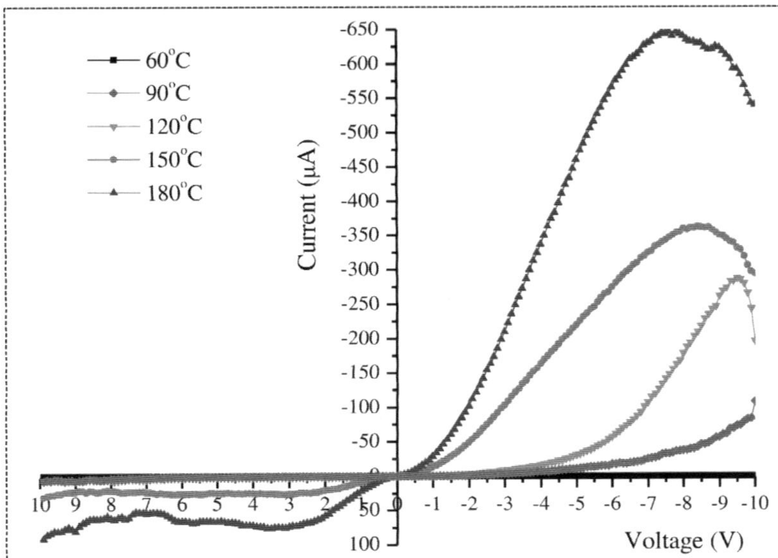

Figure 7. I-V characteristics of MIS fabricated using P3HT as semiconductor layer.

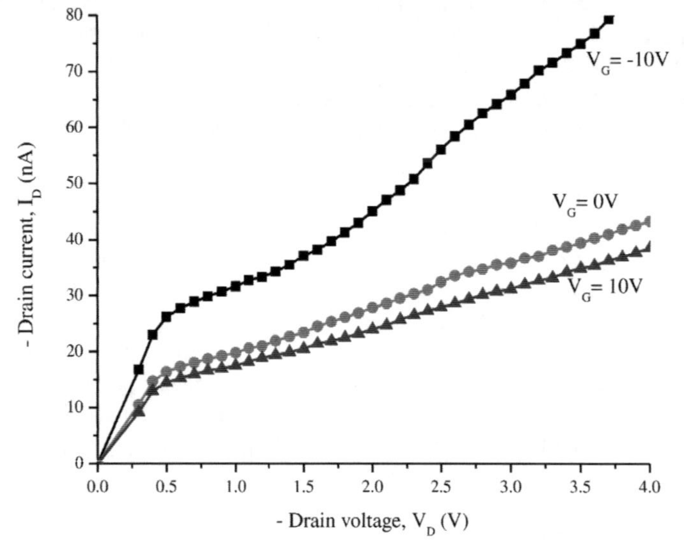

Figure 8. I_d-V_d curve characteristics curve of p-type OFET with PMMA:TiO_2 nanocomposite and P3HT as dielectric and semiconductor layer.

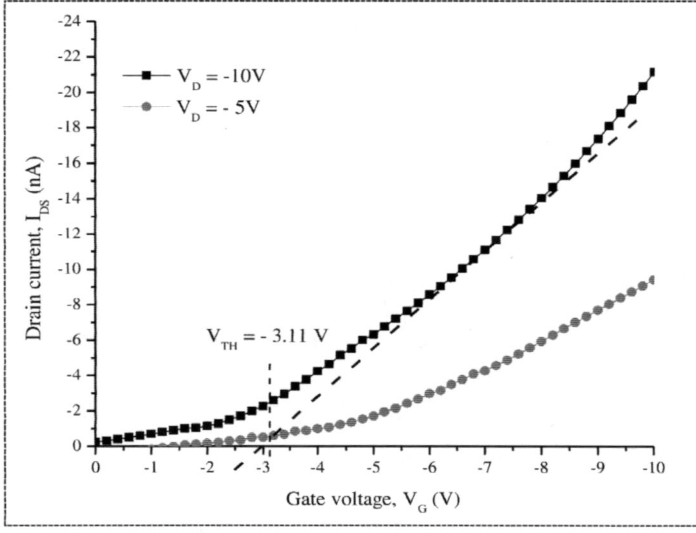

Figure 9. I_d-V_g curve characteristics curve of p-type OFET with PMMA:TiO_2 nanocomposite and P3HT as dielectric and semiconductor layer.

4. Conclusion

The fabricated MIS capacitor with different P3HT thickness and nanocomposite PMMA:TiO_2 as semiconductor and insulator, respectively on glass substrate. The results indicate that the operating voltage, V_{TH} of organic-MIS improve significantly improve when the P3HT thickness is reduce. Low surface roughness was obtained when the thickness of the P3HT film is reduce. Through its electrical properties, it was found that the insulating property of nanocomposite PMMA:TiO_2 are comparable to those usually obtained for MIS structures with other material that being use as insulator. PMMA:TiO_2 organic-inorganic films offer attractive opportunities due to its good electrical properties and an effective way to reduce the operating voltage of organic-MIS but also a useful way to investigate the interfacial capacitance that exists at the interface of semiconductor/insulator. The OFETs was subsequently fabricated and examined to identify the characteristic of organic transistor using PMMA:TiO_2 nanocomposite as dielectric layer. The fabricated p-type OFET showed a poor

IOP Conf. Series: Materials Science and Engineering **340** (2018) 012005 doi:10.1088/1757-899X/340/1/012005

performance even though the real permittivity, ε' for PMMA:TiO_2 was ~ 12, about 3 times higher than the ε' for SiO_2.

References

[1] A. Fian, A. Haase, B. Stadlober, G. Jakopic, N. B. Matsko, W. Grogger, and G. Leising, "AFM, ellipsometry, XPS and TEM on ultra thin oxide/polymer nanocomposite layers in organic thin film transistors," Analytical and Bioanalytical Chemistry, vol. 390, pp. 1455-1461, 2007.
[2] C. G. Choi and B.-S. Bae, "Organic-inorganic hybrid materials as solution processible gate insulator for organic thin film transistors," Organic Electronics, vol. 8, pp. 743-748, 2007.
[3] I. Kymissis, "OFET applications," in Organic Field Effect Transistors, A. Chandrakasan, Ed., ed: Springer US, 2009, pp. 1-6.
[4] S. Uemura, A. Komukai, R. Sakaida, T. Kawai, M. Yoshida, S. Hoshino, T. Kodzasa, and T. Kamata, "The organic FET with poly(peptide) derivatives and poly(methyl-methacrylate) gate dielectric," Synthetic Metals, vol. 153, pp. 405-408, 2005.
[5] C. R. Newman, C. D. Frisbie, D. A. d. S. Filho, J.-L. Bredas, P. C. Ewbank, and K. R. Mann, "Introduction to Organic Thin Film Transistors and Design of n-channel Organic Semiconductors," Chem. Mater, vol. 16, pp. 4436-4451, 2004.
[6] V. Subramanian, M. Toita, N. R. Ibrahim, S. J. Souri, and K. C. Saraswat, "Low-leakage germanium-seeded laterally-crystallized single-grain 100-nm TFTs for vertical integration applications," Electron Device Letters, IEEE, vol. 20, pp. 341-343, 1999.
[7] G. Crawford, Flexible flat panel displays: Wiley. com, 2005.
[8] J. T. Mabeck and G. G. Malliaras, "Chemical and biological sensors based on organic thin-film transistors," Analytical and Bioanalytical Chemistry, vol. 384, pp. 343-353, 2006.
[9] A. L. Deman and J. Tardy, "Stability of pentacene organic field effect transistors with a low-k polymer/high-k oxide two-layer gate dielectric," Materials Science and Engineering: C, vol. 26, pp. 421-426, 2006.
[10] G. Horowitz, "Organic thin film transistors: Fron theory to real devices," Journal of Material Research vol. 19, pp. 1946-1962, 2004.
[11] Y.-Y. Noh and H. Sirringhaus, "Ultra-thin polymer gate dielectrics for top-gate polymer fieldeffect transistors," Organic Electronics, vol. 10, pp. 174-180, 2009.
[12] L.N. Ismail, Z. Habibah, N. S. M. Sauki, Z. Aznilinda, S. H. Herman and M. Rusop, "Characterization of Metal-Insulator-Semiconductor Capacitor with Poly(methyl methacrylate): Titanium Dioxide as Insulator," Japanese Journal of Applied Physics, vol.52, pp.06GG02,2013.
[13] M. Morales-Acosta, M. Quevedo-López, B. Gnade, and R. Ramírez-Bon, "PMMA-SiO2 organic–inorganic hybrid films: determination of dielectric characteristics," Journal of Sol-Gel Science and Technology, vol. 58, pp. 218-224, 2010.
[14] S. Liwei, M. Liu, T. Deyu, G. Liu, L. Xinghua, and J. Zhuoyu, "Low-Voltage Organic FieldEffect Transistor With PMMA/ZrO2 Bilayer Dielectric," Electron Devices, IEEE Transactions on, vol. 56, pp. 370-376, 2009.
[15] N. B. Ukah, J. Granstrom, R. R. S. Gari, G. M. King, and S. Guha, "Low operating voltage ad stable organic field-effect transistors with poly (methyl methacrylate) gate dielectric solution deposited from a high dipole moment solvent," Applied Physics Letters, vol. 99, p. 243302, 2011.
[16] J. Robertson, "High dielectric constant gate oxides for metal oxide Si transistors," Reports on Progress in Physics, vol. 69, p. 327, 2006.
[17] M. Ahmadi, N. Phonthammachai, T. H. Shuan, T. J. White, N. Mathews, and S. G. Mhaisalkar, "Solution processable nanoparticles as high-k dielectric for organic field effect transistors," Organic Electronics, vol. 11, pp. 1660-1667.
[18] H. Wong and H. Iwai, "On the scaling issues and high-[kappa] replacement of ultrathin gate dielectrics for nanoscale MOS transistors," Microelectronic Engineering, vol. 83, pp. 1867- 1904, 2006.
[19] D. A. Buchanan, "Scaling the gate dielectric: materials, integration, and reliability," IBM J. Res.

Dev., vol. 43, pp. 245-264, 1999.
[20] Y. YoungJun, "Pentacene Based Organic Electronic Devices," PhD, Durham University, 2010.
[21] N. G. Semaltianos, "Spin coated PMMA films," Microelectronic Journal vol. 38, pp. 754 - 761, 2007.
[22] T. S. Huang, Y. K. Su, and P. C. Wang, "Poly(methyl methacrylate) Dielectric Material Applied in Organic Thin Film Transistors," Japanese Journal of Applied Physics, vol. 47, pp. 3185 - 3188, 2008.
[23] D. D. L. Chung, Functional Materials : Electrical, Dielectric, Electromagnetic, Optical and Magnetic Appliecations vol. 2. Singapore: World Scientific Publishing Co. Pte. Ltd, 2010.
[24] H. S. Nalwa, Handbook of Low and High Dielectric Constant Materials and Their Applications, Two-Volume Set: Academic Press, 1999.
[25] T. Osaka and M. Datta, Energy Storage Systems in Electronics: CRC Press, 2000.
[26] R. Arora and W. Mosch, High voltage insulation engineering: behaviour of dielectrics, their properties and applications: New Age International, 1995.
[27] T. Dakin, "Conduction and polarization mechanisms and trends in dielectric," Electrical Insulation Magazine, IEEE, vol. 22, pp. 11-28, 2006.
[28] W. Sarjeant, R. Dollinger, J. Zirnheld, F. MacDougall, and H. Goldberg, "Capacitors-past, present, and future: a trans-national perspective," in Power Modulator Symposium, 1996., Twenty-Second International, 1996, pp. 209-212.
[29] N. B. UKAH, "Low Dielectric Constant-Based Organic Field Effect Transistors and MetalInsulator-Semiconductor Capacitors," Doctor of Philoshopy, Faculty of the Graduate School, University of Missouri, 2012.
[30] J. Puigdollers, C. Voz, A. Orpella, R. Quidant, I. Martín, M. Vetter, and R. Alcubilla, "Pentacene thin-film transistors with polymeric gate dielectric," Organic Electronics, vol. 5, pp. 67-71, 2004.
[31] S. P. Tiwari, V. R. Rao, T. Huei Shaun, E. B. Namdas, and S. G. Mhaisalkar, "Pentacene Organic Field Effect Transistors on Flexible substrates with polymer dielectrics," in VLSI Technology, Systems and Applications, 2007. VLSI-TSA 2007. International Symposium on, 2007, pp. 1-2.
[32] J.-S. Lim, P.-K. Shin, B.-J. Lee, and S. Lee, "Plasma polymerized methyl methacrylate gate dielectric for organic thin-film transistors," Organic Electronics, vol. 11, pp. 951-954, 2010.
[33] A. Facchetti, M. H. Yoon, and T. J. Marks, "Gate dielectrics for organic field-effect transistors: New opportunities for organic electronics," Advanced Materials, vol. 17, pp. 1705-1725, 2005.
[34] M. D. Morales-Acosta, C. G. Alvarado-Beltran, M. A. Quevedo-Lopez, B. E. Gnade, A. Mendoza-Galvan, and R. Ramirez-Bon, "Adjustable structural, optical and dielectric characteristics in sol-gel PMMA-SiO2 hybrid films," Journal of Non-Crystalline Solids, vol. 362, pp. 124-135, 2013.

Acknowledgments
Authors are grateful to the Research Management Unit (RMU), Universiti Teknologi MARA for providing us the RAGS grant (600-RMI/RAGS 5/3 (205/2014)). The authors also would like to acknowledge NANO-SciTech Centre, UiTM Shah Alam, for the use their AFM equipment.

IOP Conf. Series: Materials Science and Engineering **340** (2018) 012006 doi:10.1088/1757-899X/340/1/012006

Dilute electrodeposition of TiO_2 and ZnO thin film memristors on Cu substrate

F B Fauzi[1], M H Ani[1], S H Herman[2], M A Mohamed[3]

[1] Department of Manufacturing and Materials Engineering, International Islamic University Malaysia (IIUM), Jalan Gombak, 53100 Kuala Lumpur, Malaysia
[2] NANO-Electronic Centre (NET), Faculty of Electrical Engineering, Universiti Teknologi MARA, 40450 Shah Alam, Selangor, Malaysia
[3] Institute of Microengineering and Nanoelectronics, Universiti Kebangsaan Malaysia, 43600 UKM-Bangi, Selangor, Malaysia

Email: mhanafi@iium.edu.my

Abstract. Memristor has become one of the alternatives to replace the current memory technologies. Fabrication of titanium dioxide, TiO_2 memristor has been extensively studied by using various deposition methods. However, recently more researches have been done to explore the compatibility of other transition metal oxide, TMO such as zinc oxide, ZnO to be used as the active layer of the memristor. This paper highlights the simple and easy-control electrodeposition to deposit titanium, Ti and zinc, Zn thin film at room temperature and subsequent thermal oxidation at 600 °C. Gold, Au was then sputtered as top electrode to create metal-insulator-metal, MIM sandwich of Au/TiO_2-Cu_2O-CuO/Cu and Au/ZnO-Cu_2O-CuO/Cu memristors. The structural, morphological and memristive properties were characterized using Field Emission Scanning Electron Microscopy, FESEM, X-Ray Diffraction, XRD and current-voltage, *I-V* measurement. Both Au/TiO_2-Cu_2O-CuO/Cu and Au/ZnO-Cu_2O-CuO/Cu memristivity were identified by the pinched hysteresis loop with resistive ratio of 1.2 and 1.08 respectively. Empirical study on diffusivity of Ti^{4+}, Zn^{2+} and O^{2-} ions in both metal oxides show that the metal vacancies were formed, thus giving rise to its memristivity. The electrodeposited Au/TiO_2-Cu_2O-CuO/Cu and Au/ZnO-Cu_2O-CuO/Cu memristors demonstrate comparable performances to previous studies using other methods.

1. Introduction

Memristor is a promising alternative for the next generation non-volatile memory system. The limitations on size and manufacturing cost of the current memory storage especially for dynamic random access memory, DRAM has closely reached their maximum limit and has driven a tremendous amount of research on memristor. Memristor was first discovered as the fourth fundamental element by Chua [1] in 1971 and it was experimentally demonstrated in 2008 by William et al. [2]. Memristor is a non-volatile memory which is also known as a resistive random access memory, RRAM made from a simple structure of MIM sandwich. The advantages of memristor are its fast switching speed, low power consumption and high density [1]. The uniqueness of the memristor is that it will remember the most recent resistance even when the voltage is off [3]. Previous researches discovered that transition metal oxides, TMO such as TiO_2 [3]-[6], ZnO [7]–[9], nickel (II) oxide, NiO [10], perovskite oxides [11] and cuprous oxide, Cu_2O [12] have the ability to create memristive effects when used as the intermediate active layer of memristor.

The most popular material used in memristor is TiO_2 due to its simple structure and compatibility with the Complementary Metal-Oxide-Semiconductor, CMOS manufacturing process [4]. Ti is an expensive material and very complicated to handle due to its reactivity to oxygen. In recent years, many methods have been used to fabricate memristive devices. So far the most common methods used have been radio frequency, RF sputtering [13]–[19], thermal oxidation [4], [7], chemical vapor deposition, CVD [20], [21], pulsed laser deposition, PLD [22], sol gel [6], [23], atomic layer deposition, ALD [24] and electrodeposition [7], [8], [25]. Recently, ZnO has attracted a lot of attention due to its non-toxicity and prevalence in semiconductor material for electronic devices applications such as light emitting diodes, LED, photodetectors, sensors, and solar cells.

In this paper, thin films of TiO_2 and ZnO were fabricated on copper, Cu substrate by using electrodeposition and thermal oxidation methods that yields comparable results to the expensive, complex and time consuming existing methods. Cu substrate has been chosen due to its resistance to corrosion, high electrical conductivity and its wide use in electronic devices. The memristive performance of Au/TiO_2-Cu_2O-CuO/Cu and Au/ZnO-Cu_2O-CuO/Cu samples were compared and evaluated.

2. Experimental procedures

The electrodeposition setup consists of working electrode and counter electrode placed in a dilute solution of 0.06 M titanium trichloride, $TiCl_3$ and 0.005 M zinc chloride, $ZnCl_2$ electrolytic baths. The supporting precursor hydrogen peroxide, H_2O_2 was added to the $TiCl_3$ to pre-oxidize Ti^{3+} to Ti^{4+}. $TiCl_3$ electrolyte bath was prepared as reported by Chang et al. [26] by adding 0.075 M of hydrochloric acid followed by addition of H_2O_2. 1 M $ZnCl_2$ stock solution was made by dissolving 13.65 g of $ZnCl_2$ from Ajax Finechem in 100 ml distilled water. Then, it was diluted to 0.005 M. A fixed voltage of 1 V was used during the electrodeposition process for Zn deposition, but a higher voltage of 10 V was needed for Ti deposition. The acrylic cell holder was used to clamp the plates with a fixed spacing, *d* of 20 mm and deposition diameter, *a* of 12 mm shown in Figure 1 (a) and schematic diagram of the electrodeposition setup shown in Figure 1 (b).

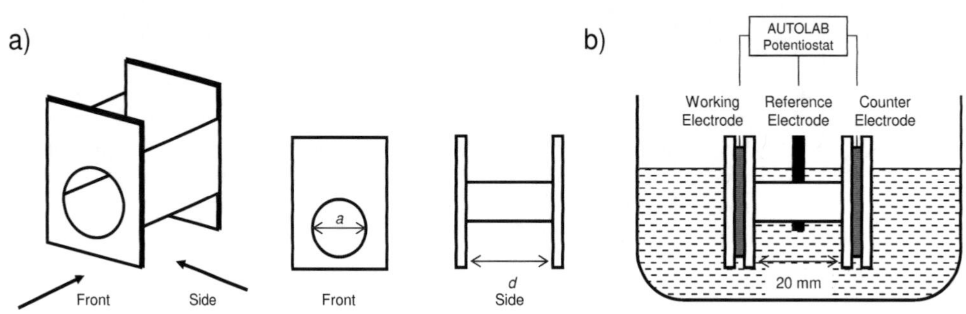

Figure 1. Schematic diagram for electrolytic cell holder (a) and electrodeposition setup (b).

The working electrode was Cu foil for both cases, while the counter electrode was Ti plate for Ti/Cu sample and Zn foil for Zn/Cu sample. Both thin films were deposited for 1 minute deposition time at its respective voltage. After the deposition, the samples underwent thermal oxidation process at a temperature of 600 °C for 60 minutes [27]. It was then coated with Au to form a simple MIM junction by using an auto fine coater (JEOL JFC-1600). Both deposited samples then underwent *I-V* measurements by applying DC voltage from -2 V to 2 V with a step voltage of 0.1V. FESEM (JEOL JSM-6700F), XRD (SHIMADZU XRD-6000) and potentiostat (PGSTAT302N, AUTOLAB) were used to characterize the thin films for surface morphology, structure and *I-V* measurement. All experiments and measurements were performed at room temperature.

IOP Conf. Series: Materials Science and Engineering **340** (2018) 012006 doi:10.1088/1757-899X/340/1/012006

3. Results and Discussion

The XRD patterns of both thin films after thermal oxidation were illustrated in Figure 2 (a). Miller indices were included at each diffraction peak. Deposited Ti and Zn were fully oxidized to TiO_2 and ZnO. TiO_2 peaks were observed at 25.21° (101), 49.02° (200) and 61.21° (118) conforming to JCPDS Card 21-1272. ZnO peaks also matches the JCPDS Card 36-1451 at 2θ of 35.68° (101) and 44.26° (102). The existence of CuO (JCPDS Card 80-1917) and Cu_2O (JCPDS Card 5-0667) diffraction peaks indicated that the Cu (JCPDS Card 85-1326) was being oxidized as show in Figure 2 (b). Hence, not only TiO_2 and ZnO but also copper oxides were also possibly functioning as the active layer for the memristor.

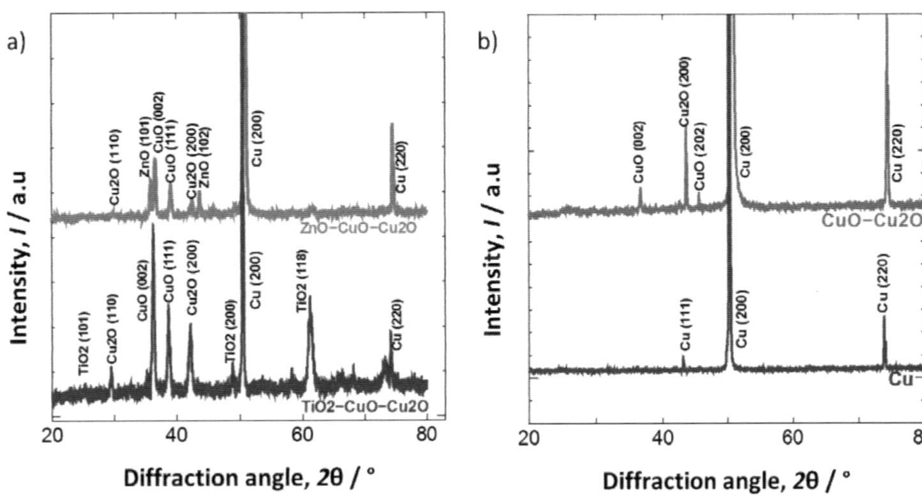

Figure 2. XRD results of deposited thin films after thermal oxidation (a) and Bare Cu substrate before and after thermal oxidation (b).

As observed, there were peaks shifted to higher 2θ after thermal oxidation process. Thermal oxidation has resulted in the Cu peaks becoming broader and the intensity was increased. Their peaks also were shifted to a higher angle. This indicates the formation of copper oxides and at the same time inter-diffusion between atoms has occurred. The phase formation during thermal oxidation occurred through solid-state diffusion of ions. The 2θ values of Cu diffraction peaks corresponding to (111), (200) and (220) had shifted towards higher angle value thus indicating the reduction of the lattice parameter of Cu. This is due to the diffusion of Cu which during the formation of oxides. This lattice changes can be calculated by using Bragg's Law in equation (3.1).

$$n\lambda = 2d \sin\theta \tag{3.1}$$

where n is equal to 1, λ is the wavelength of the x-ray, d is the lattice distance and θ is the incident angle. The lattice distances, *d* for corresponding (111), (200) and (220) after the thermal oxidation process was reduced by 2.5 x 10^{-3} nm, 1.1 x 10^{-3} nm, and 6.0 x 10^{-4} nm respectively. From here, the shifted peak shows the reduction in lattice parameter of Cu which happened during the heat treatment where diffusion of ions occurred. Figure **3** shows the FESEM micrographs for surface morphologies of (a) TiO_2-Cu_2O-CuO and (b) ZnO-Cu_2O-CuO at high magnification (x60k). The microstructure of TiO_2-Cu_2O-CuO exhibits granular grains in contrast to ZnO-Cu_2O-CuO that displays a rod-like structure with similar length and diameter.

Figure 3. Surface morphology of synthesized TiO_2-Cu_2O-CuO/Cu (a) and ZnO-Cu_2O-CuO/Cu (b) after thermal oxidation at x60k magnification.

The *I-V* measurements of both samples show the pinched hysteresis loops as shown in Figure 4 (a) and (b) which follows the characteristics of a memristor. Higher maximum and minimum current value for Au/ZnO-Cu_2O-CuO/Cu was bigger than the Au/TiO_2-Cu_2O-CuO/Cu sample. The maximum and minimum current measured were 52 μA, -61 μA for Au/TiO_2-Cu_2O-CuO/Cu and 93 μA, -103 μA for Au/ZnO-Cu_2O-CuO/Cu respectively.

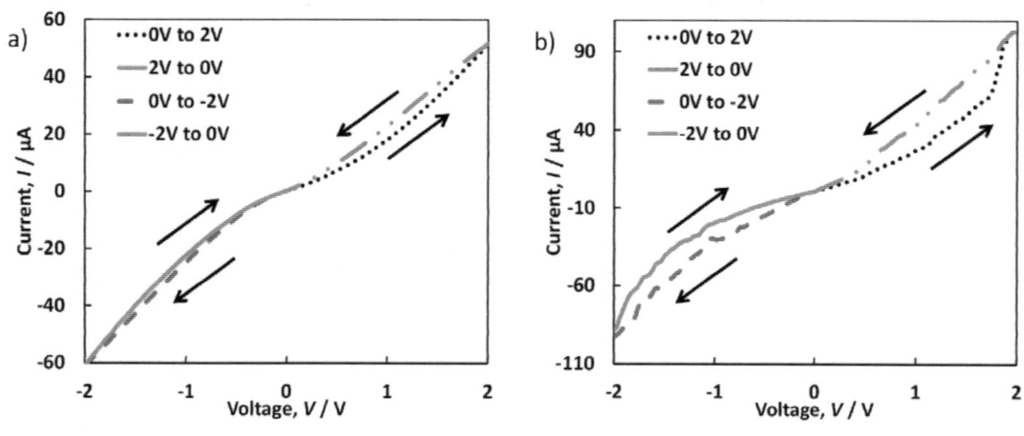

Figure 4. *I-V* hysteresis loops from synthesized Au/TiO_2-Cu_2O-CuO/Cu (a) and Au/ZnO-Cu_2O-CuO/Cu (b) at 1 minute deposition time.

The polarity dependence of the resistance to voltage is called bipolar resistive switching [11]. Resistive switching behavior was illustrated in

Figure **4** where the current changes from high resistance state, HRS to low resistance state, LRS at a set voltage from '0 V to 2 V' path to '2 V to 0 V' path. The difference of HRS and LRS creates the hysteresis loop in *I-V* curve. Calculated HRS and LRS of both samples were tabulates in Table 1. Au/TiO_2-Cu_2O-CuO/Cu gave greater difference in between HRS and LRS of 8.83 kΩ with HRS/LRS ratio of 1.2. While Au/ZnO-Cu_2O-CuO/Cu have HRS and LRS difference of 2.02 kΩ with HRS/LRS ratio of 1.08.

Table 1. Calculated HRS and LRS.

Sample	ΔI_{max} (μA)	HRS (kΩ)	LRS (kΩ)	HRS – LRS (kΩ)	HRS/LRS
Au/ZnO-Cu_2O-CuO/Cu	21.0	27.05	25.0	2.02	1.08
Au/TiO_2-Cu_2O-CuO/Cu	5.0	50.7	41.9	8.83	1.2

HRS/LRS ratio of Au/ZnO-Cu_2O-CuO/Cu shows results that is within the range of our previous findings reported by Marmezee et al. [7]. Electrodeposition is suitable to be used as a fabrication method because of its reproducibility and comparable results to other studies reported by other groups.

The difference between HRS and LRS in memristor occurred because of the diffusion of defect vacancy in oxide layers when polarized. The formation vacancy defects during oxidation in a metal oxide depends on the species diffusivity according to Fick's first law of diffusion (steady-state) in equation (3.2):

$$J = -D\,\frac{\partial c}{\partial x} \qquad (3.2)$$

where J is the flux, D is the diffusivity coefficient, c is the concentration of each species and x is the displacement (length of diffusion).

From experimental data on oxygen ion, O^{2-} diffusivity in ZnO compiled by Erhart and Albe [28], [29], it can be postulated that at 600 °C the diffusivity of O^{2-} in ZnO is around 10^{-22} $cm^2\,s^{-1}$ and zinc ions, Zn^{2+} in ZnO is about 10^{-18} $cm^2\,s^{-1}$. It shows that the Zn^{2+} has greater diffusivity compared to O^{2-} in ZnO. For TiO_2, the temperature dependence of self-diffusion coefficient for titanium (IV) ions, Ti^{4+} and O^{2-} in TiO_2 are compiled by Pereloma et al. [30] and Nowotny [31]. It is postulated that at 600 °C the diffusivity of O^{2-} in TiO_2 is around 10^{-24} $cm^2\,s^{-1}$ and Ti^{4+} in TiO_2 is about 10^{-19} $cm^2\,s^{-1}$. It shows that the Ti^{4+} has greater diffusivity compared to O^{2-} in TiO_2. The order of diffusivity of each species are as follows in (3.3):

$$D_{Zn^{2+}\ in\ ZnO} > D_{Ti^{4+}\ in\ TiO_2} > D_{O^{2-}\ in\ ZnO} > D_{O^{2-}\ in\ TiO_2} \qquad (3.3)$$

The diffusivity of cation possesses greater value than the diffusivity of anion by a factor of $10^3 \sim 10^4$ which creates metal vacancies instead of oxygen vacancies. The formation of metal vacancy defects in transition metal oxide produces the pinched hysteresis loop in memristor. We conclude that the ability of memristor to be used as memory device to store data relies on the existence of metal vacancy defects come from different diffusivities in the metal oxide active layer.

4. Conclusion

In summary, it can be demonstrate that TiO_2-Cu_2O-CuO and ZnO-Cu_2O-CuO thin films are successfully deposited using ultra-dilute solution on Cu substrates through the electrodeposition and thermal oxidation method. Both samples showed the hysteresis loop in the *I-V* measurements where Au/TiO_2-Cu_2O-CuO/Cu has bigger HRS-LRS value of 8.83 kΩ compared to Au/ZnO-Cu_2O-CuO/Cu of 2.02 kΩ. However based on the HRS/LRS ratio, both Au/TiO_2-Cu_2O-CuO/Cu and Au/ZnO-Cu_2O-CuO/Cu have comparable memristive performances of 1.2 and 1.08 respectively. From the postulation of diffusivity, the memristive effects occurred due to the formation of metal vacancy defects as the diffusion of each ions are $D_{Zn^{2+}\ in\ ZnO} > D_{Ti^{4+}\ in\ TiO_2} > D_{O^{2-}\ in\ ZnO} > D_{O^{2-}\ in\ TiO_2}$. The diffusivity of cation greater than anion in $10^3 \sim 10^4$ factor.

Acknowledgement
This work was financially supported by the Niche Research Grant Scheme, Ministry of Education Malaysia (600-RMI/NRGS 5/3 (7/2013)) and Research Acculturation Grant Scheme (RAGS13-002-0065), Universiti Teknologi MARA (UiTM) and International Islamic University Malaysia (IIUM).

References

[1] Chua L O 1971 *IEEE Transations Circuit Theory* **CT-18** 507–519
[2] Strukov D B, Snider G S, Stewart D R, and Williams R S 2008 *Nature* **453** 80–83
[3] Williams R S 2008 *IEEE Spectr.* **45** 28–35
[4] Li Y T *et al* 2010 *10th IEEE International Conference on Solid-State and Integrated Circuit Technology* 1148–1150
[5] Yang J J *et al* 2011 *Applied Physics A* **102** 785–789
[6] Gale E, Mayne R, Adamatzky A and Costello B L 2014 *Materials Chemistry and Physics* **143** 524–529
[7] Yusoff M M, Ani M H and Suryanto 2013 *Advance Materials Research* **701** 172–175
[8] Fauzi F B, Othman R, Mohamed M A, Herman S H, Ahmad Azhar A Z and Ani M H 2015 *Materials Transactions* **56** 1302-1306
[9] Muhammad N M, Duraisamy N, Rahman K, Hyun W D, Jeongdae J and Kyung H C 2013 *Current Applied Physics* **13** 90–96
[10] Zhang R, Yuldashev S, Lee J C, Yalishev V, Kang T W and Fu D J 2013 *Microelectronic Engineering* **112** 31–34
[11] Sawa A 2008 *Materials Today* **11** 28–36
[12] Li Y, Long S, Liu Q, Lu H, Liu S and Liu M 2011 *Chinese Science Bulletin* **56** 3072–3078
[13] Yakopcic C, Shin E, Taha T M, Subramanyam G, Murray P T and Rogers S 2010 *The* 2010 *International Joint Conference on Neural Networks (IJCNN)* (Barcelona) 1-4
[14] Kim S, Jeong H Y, Kim S K, Choi S Y and Lee K J 2011 *Nano Letters* **11** 5438–5442
[15] Kumar A, Rawal Y and Baghini M S 2012 *2012 International Conference on Emerging Electronics* 1–3
[16] Prodromakis T, Michelakis K and Toumazou C 2010 *Proceedings of 2010 IEEE International Symposium on Circuits and Systems* (Paris) **2** 1520–1522
[17] Yang J J, Pickett M D, Li X, Ohlberg D A A, Stewart D R and Williams R S 2008 *Nature Nanotechnology* **3** 429–433
[18] Pickett M D *et al* 2009 *Journal of Applied Physics* **106** 1–6
[19] Rosli A B, Marbie M M, Herman S H and Ani M H 2014 *Journal of Nanomaterials* **2014** 1–7
[20] Asafa T B, Tabet N and Said S A M 2013 *Neurocomputing* **106** 86–94
[21] Schindler C, Thermadam S C P, Waser R and Kozicki M N 2007 *IEEE Transactions on Electron Devices* **54** 2762–2768
[22] Jia C H, Dong Q C and Zhang W F 2012 *Journal of Alloys and Compounds* **520** 250–254
[23] Gergel-Hackett N *et al* 2009 *IEEE Electron Device Letters* **30** 706-708
[24] Kwon D H *et al* 2010 *Nature Nanotechnology* **5** 148–153
[25] Fauzi F B, Ani M H, Othman R, Ahmad Azhar A Z, Mohamed M A and Herman S H 2015 *IOP Conference Series: Materials Science and Engineering* **99** 12002
[26] Hu C C, Hsu H C and Chang K H 2012 *Journal of the Electrochemical Society* **159** D418–D424
[27] Khanlary M R, Vahedi V and Reyhani A 2012 *Molecules* **17** 5021–5029
[28] Erhart P and Albe K 2006 *Physical Review B* **73** 1–9
[29] Erhart P and Albe K 2006 *Applied Physics Letters* **88** 1-3
[30] Pereloma E , Savvakin D, Carman A, Gazder A and Ivasishin O 2012 *Key Engineering Materials* **520** 49–56
[31] Nowotny J 2012 *Oxide semiconductors for solar energy conversion—Titanium dioxide* (Boca Raton, CRC Press)

IOP Conf. Series: Materials Science and Engineering **340** (2018) 012007 doi:10.1088/1757-899X/340/1/012007

Investigation of the effect of Anodized Duration toward Photocatalytic Performance of Nb_2O_5

Ahmad Sabirin Zoolfakar[1], Nurul Atiqah Mokhtar[1], Rozina Abdul Rani[2], Nur Samihah Khairir[1], Mahzaton Aqma Abu Talip[1], Mohamad Hafiz Mamat[1], Rosmalini Abdul Kadir[1], M. Rusop[2]

[1]Faculty of Electrical Engineering, University Teknologi Mara, 40450 Shah Alam, Selangor, Malaysia

[2]NANO-SciTech Centre, Institute of Science, Universiti Teknologi MARA, 40450, Shah Alam, Selangor, Malaysia

Email: ahmad074@salam.uitm.edu.my

Abstract. Highly oriented Nb_2O_5 nanoporous network produced *via* anodization for photocatalytic activity of methyl orange (MO) is presented. The anodization duration was varies from 0.5 to 2 hours and the photocatalytic performance is observed by degradation of MO solution. The Nb_2O_5 nanoparticles were added in MO solution and were exposed to the solar simulator for 3 hours. The morphology of Nb_2O_5 nanoporous and the photocatalytic performance are characterized in Field Emission Scanning Electron Microscopy (FESEM) and UV-Vis spectrophotometer, respectively. The result shows that different duration of anodized produce different sizes of nanoporous diameter that will significantly affect the photocatalytic performance. The 1.5 hours of anodized has the largest diameter size of nanoporous and exhibited the best photocatalytic performance

1. Introduction

Wastewater is one of the main issues in Malaysia due to the water contamination produced by the industries that produce leather, textiles, plastic and other dying industries [1]. Dyes have negative impact to many types of life for both esthetical and toxicological reasons [1]. Dyes used in industries for various field namely textile to color their products. These industries produced large amount of colored wastewater and will caused water pollution. It is estimated from 40,000 to 50,000 tons of dye are wasted to surface water every year [1].

Public awareness towards these type of pollution have caused the scientists to develop numerous treatment technologies for dyes cure such as adsorption, filtration, photocatalytic degradation and many more [2]. One of the treatment technologies that has obtained huge attention is photocatalytic because its speed, high efficiency and no side pollution [3, 4].

The photocatalytic using niobium pentoxide (Nb_2O_5) is considering as favourable material because of its chemical stability and environmental friendly [5]. It is an important semiconductor with a bandgap of ~3.1 eV and has been discovered to potentially serve in various applications such as photocatalysis, catalysis and sensors [6, 7]. In this research, the Nb_2O_5 nanoparticles were used instead of the thin film. The nanoparticles of Nb_2O_5 are obtained after anodized of Nb_2O_5 thin film. Anodized is a commonly used method in nano-fabrication because of its capability to produce highly porous and ordered oxide morphologies [8, 9]. Usually the anodized process includes two electrodes which are cathode (negative electrode) and anode (positive electrode) [10]. Both of the electrodes are both immersed in the liquid electrolyte and when voltage or electrical current is turned on, the electrochemical reactions will take place on the surface of anode electrode [11].

There are many research have been published on investigating factors that affect photocatalytic performance but no research has been focus on the effect of anodized duration on photocatalytic of Nb2O5 nanoparticles performances. Therefore, it is necessary to investigate this parameter in order to treat the dye wastewater. The aim of this research is to investigate the effect of anodized duration toward photocatalytic performance of Nb_2O_5 nanoparticles. Variation of anodized duration caused change in size of Nb_2O_5 nanoporous that will affect the photocatalytic performance of Nb_2O_5 nanoparticles. Thus, the photocatalytic performance of the Nb_2O_5 nanoparticles is determined with degradation of methyl orange solution that is observed by the value of transmittance during photocatalytic activity.

2. Methodology

2.1. Fabrication of Nb_2O_5 nanoparticles

The niobium foil with a thickness of 0.25 mm was cut into smaller pieces 1.0 cm X 2.5 cm. The surface of each sample was cleaned ultrasonically with acetone, ethanol, distilled water and blow with nitrogen gas to avoid any contaminates. The electrolyte for anodized was prepared with 0.5g NH_4F powder, 50 ml ethyl glycol, and 4% from ethyl glycol of distilled water (2 ml). The electrolyte was stirred on the hot plate and kept the temperature of 50℃. Two electrodes of cathode (platinum) and anode (Nb_2O_5 thin film) was set up as shown in Figure 1. The anodized duration was varies to 0.5, 1.0, 1.5 and 2.0 hours. Thus, four samples of Nb_2O_5 thin films with different duration of anodized were obtained. The samples were annealed after the anodized with temperature 440℃ for 20 minutes with steady ramp up and ramp down of 2℃/min to improve crystallinity. After that, the Nb_2O_5 nanoparticles were prepared by using a cleaned razor blade to scrape the oxides nanostructures from the Nb foil substrates. The flow chart of the anodization of Nb_2O_5 steps have been summarize in Figure 2.

2.2. Photocatalytic Process

0.01 g Methyl orange powder was stirred in 100ml of distilled water until dissolved. The 5 milligrams mass Nb_2O_5 nanoparticles was added and stirred. Then the solution was put in a dark solar simulator for 1 hour to ensure that the solution reached equilibrium state. Then, the solar light in solar simulator turned on. The process take placed for 3 hours, but every half hour, the solution was taken out for UV-Vis to observe the change of its decolorization. All of the steps for photocatalytic activity have been summarize in the Figure 3.

2.3. Photocatalytic characterization

In every 30 minutes of the photocatalytic activity in the solar simulator, the solution was removed and observed using UV-Vis equipment. The original solution before photocatalytic process was put first to set as the baseline. The aim of this characterization is to see the reaction of Nb_2O_5 nanoparticles with the methyl orange solution. The transmission data was observed.

3. Result and Discussion.

3.1. Effect of duration of anodized to the Nb_2O_5 nanoporous size.

The anodized Nb_2O_5 layer is formed by transferring a direct current through electrolytic solution. At the cathode which is the negative electrode, hydrogen is formed. Together with hydrogen and oxygen that formed at the Nb anode, a build-up Nb_2O_5 is created. The electrolyte used in anodized process contains fluorine ions and the chemical dissolution of fluorine ions happened during anodized process [12]. When 10 V of power supply connects, the reaction of chemical dissolutions takes place in Nb_2O_5 thin film that acts as an anode. Thus the nanoporous produced in the Nb_2O_5 thin films.

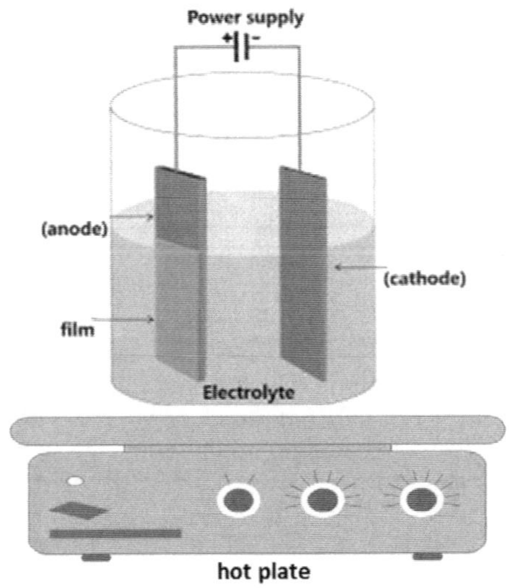

Figure 1. Illustration of anodization set up.

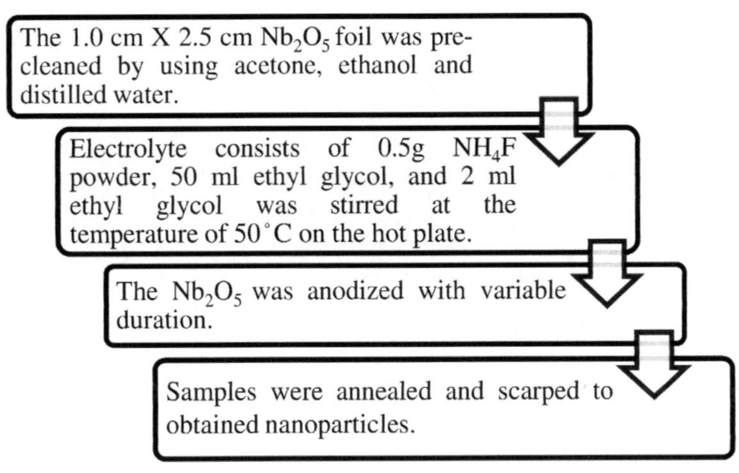

Figure 2. Flow chart of Nb_2O_5 anodization process.

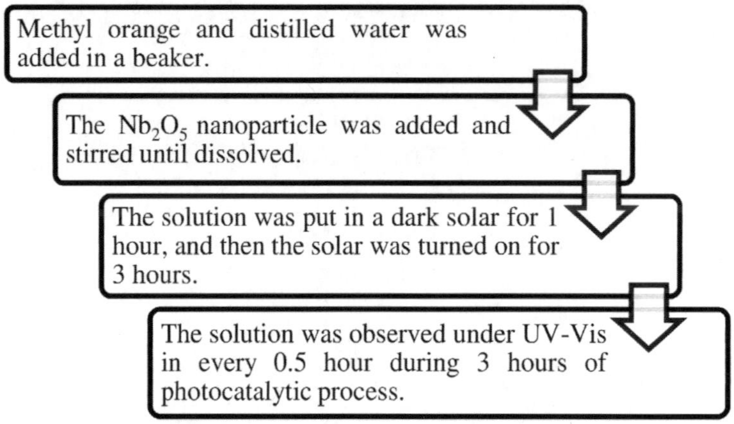

Figure 3. Flow chart of photocatalytic process

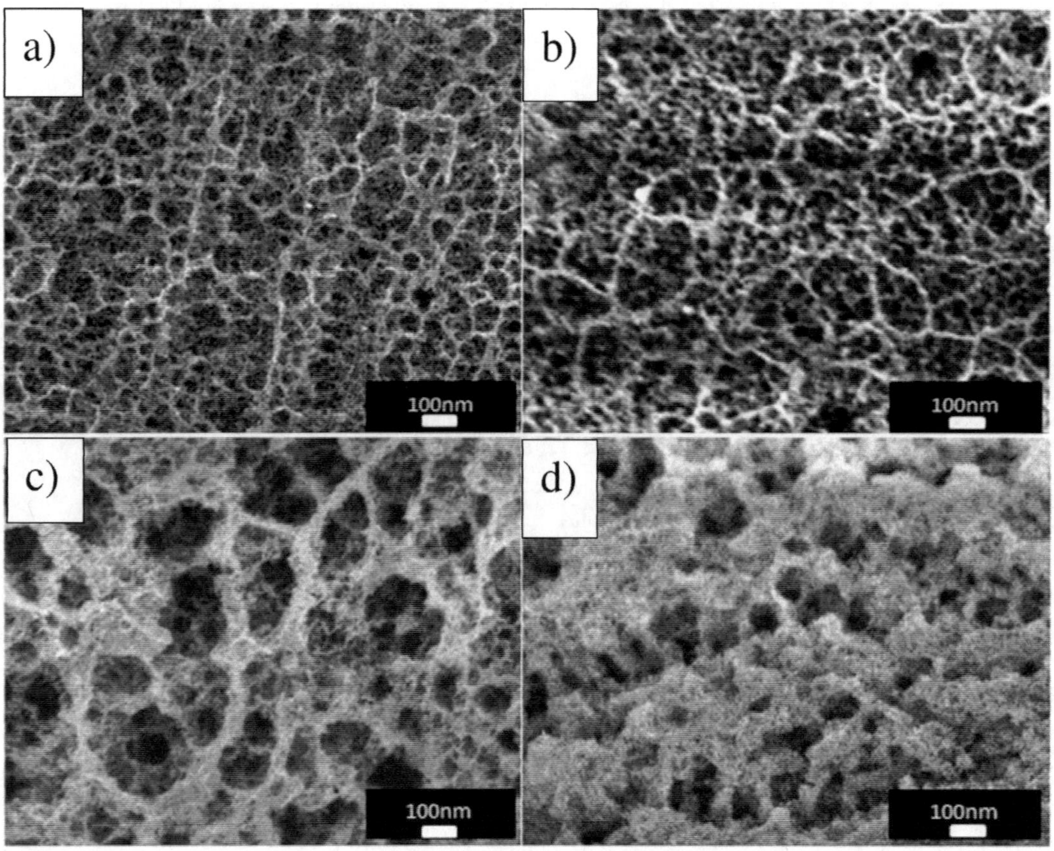

Figure 4. FESEM images of Nb_2O_5 nanoporous prepared in different time of anodized, (a) top view of anodized Nb_2O_5 at 0.5h, (b) Nb_2O_5 at 1.0h, (c) Nb_2O_5 at 1.5h, (d) Nb_2O_5 at 2h.

Figure 4 shows the FESEM images that show the surface morphology of the anodized Nb_2O_5 nanoparticles. The top view shows the structure of the Nb_2O_5 nanoparticles. As shown in the figure, a highly uniform pore distribution can be seen with nanosized pores. The sizes of nanoporous are depending on duration of anodized. The longer duration caused more chemical dissolution on the thin film and size of the nanoporous will increase until it reaches the optimum period of anodization. The size of nanoparticles for anodized in 1.5 hours is the largest compared to 0.5, 1.0 and 2.0 hour anodized. 1.5 hour of anodized is the optimum duration where the size of nanoporous is the largest and highly oriented nanoporous. However, the sample when undergo 2 hours anodized shows that the film is partially ruptured. When the anodized duration exceed more than 1.5 hours, the chemical dissolution of fluoride ions in the electrolyte still keep going and it will become excessive as the time increase. Fluoride ions have a crucial part on the surface morphology of the porous. It has been reported by *Gong et al.* that the fluoride ion concentration changes the formation and morphology of titania samples after undergo anodization process [12].

The concept can also be applied to niobium which is the dissolution of fluorine ions affect the formation of morphology. The oxide layer will fail to form uniformed pores if the quantity of fluorine ions in the electrolyte is insufficient [12]. In this case, the quantity of fluorine ions have become excessive .This situation has caused the thin film's surface become partially ruptured and the size of nanoporous no longer increase. The anodized duration after 1.5 hours does not effective anymore. It can be observed that the 1.5 hours duration of anodization is the optimum condition to obtain the largest size of nanoporous.

3.2. Effect of Nb_2O_5 nanoparticles' size to the photocatalytic performance.

There are so many factors that can affect the photocatalytic performance and one of the factors is the absorption surface area of material [13]. Absorption surface areas which are the nanoporous play important role in photocatalytic because the absorption and diffusion of molecules occur at the nanoporous [13]. The larger sizes of nanoporous increase the efficiency of absorption and diffusion of molecules of dye which is the methyl orange solution so that a better photocatalytic performance can be achieved. The photocatalytic process is done in a solar simulator by exposing the solution with solar light with the range of 400 to 800 nm wavelength. Before the solar light turn on, the solution must keep in the dark to ensure the absorption-desorption equilibrium. The photocatalytic performance in this research is shown by measuring their transmittance. Transmittance is a measurement quantity of light absorb by a substance [13]. If a sample has 100% transmittance means there are no light absorption occurs. The increase of light absorption will cause the decrease in transmittance.

Figures 5, 6, and 7 are representing the transmittance of sunlight over wavelength for different periods of photocatalytic activity for the sample that undergoes anodization period for 0.5 hour, 1.5 hour and 2 hours respectively. In every 30 minutes of photocatalytic activity, the methyl orange solution is characterized by using UV-Vis spectra to observe the change of transmittance.

Figure 8 shows the comparison of magnitude transmittance percentage for different duration anodized of Nb_2O_5 nanoparticles. This figure shows that the photocatalytic performance increases from 0.5 and 1.0 hour of anodized. The larger size of nanoparticles means the larger absorption surface area toward methyl orange ion [14]. The larger absorption of surface area can absorb more negative charge (MO^-) comparing to the small surface area in the same time interval. The larger sizes of nanoporous increase the efficiency of absorption and diffusion of molecules of methyl orange and give the better photocatalytic performance. In contrast, the performances for Nb_2O_5 that undergo anodized for 2 hours become the worst. This happen because the structure of Nb_2O_5 nanoparticles is ruptured. So the reaction between Nb_2O_5 powders with methyl orange already disturbed and cannot absorb many negative charge of methyl orange. These results has been supported with the surface morphology of these Nb_2O_5 nanoparticles which is 2 hours anodized sample has the worst morphology compared to others. Thus it can be said that, the nanoporous sizes really affect the photocatalytic performance. The best size of nanoporous which is the largest, give the best performance of photocatalytic as it has the largest absorption surface area to absorb more negative charge (MO^-) during photocatalytic activity.

Based on the figures 5, 6, and 7, one should observe that the degradation of methyl orange increase as the time increase. The degradation of methyl orange can occurs because surface charges of Nb_2O_5

IOP Conf. Series: Materials Science and Engineering **340** (2018) 012007 doi:10.1088/1757-899X/340/1/012007

absorb the negative charge (MO^-) during photocatalytic activity. As the duration of photocatalytic increase, more negative charge (MO^-) can be absorbed by Nb_2O_5 nanoparticles. As a result the photocatalytic performances improve. The negative charges (MO^-) can be absorbed because of the excited electron that gain energy from solar light and the excited electron be absorbed into the nanoporous of Nb_2O_5. The longer duration of photocatalytic process that exposed the methyl orange solution to the solar light gives more energy to the electron and caused surface charges of Nb_2O_5 absorb more negative charge (MO^-). The transmittances become lower shows the increase quantities of light absorb by the methyl orange when duration of photocatalytic become longer that will result in better photocatalytic performance as said above.

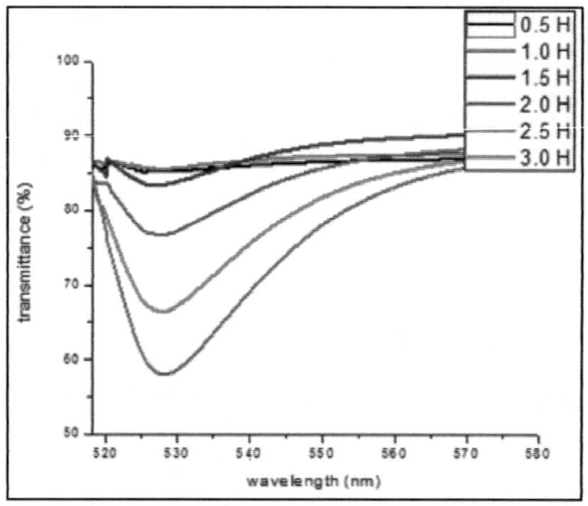

Figure 5. Transmittance of methyl orange for 0.5 H anodized of Nb_2O_5 nanoparticles.

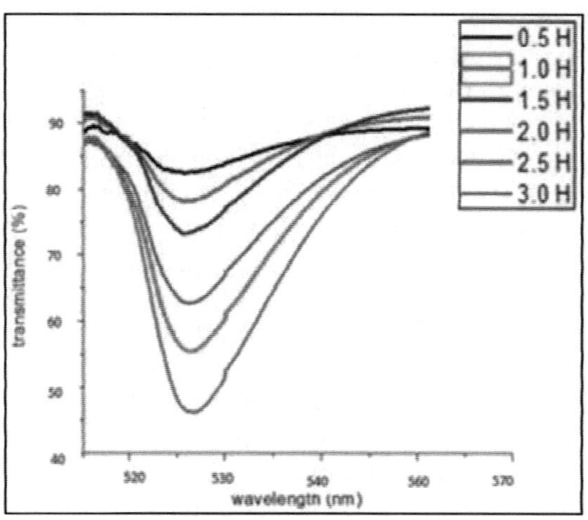

Figure 6. Transmittance of methyl orange for 1.5 H anodized of Nb_2O_5 nanoparticles.

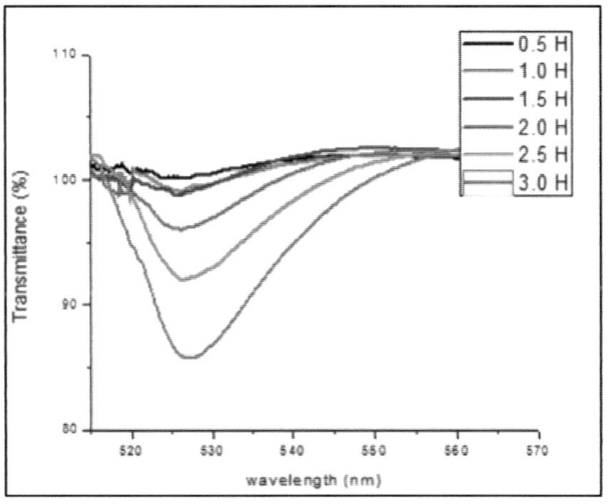

Figure 7. Transmittance of methyl orange for 2.0 H anodized of Nb_2O_5 nanoparticles.

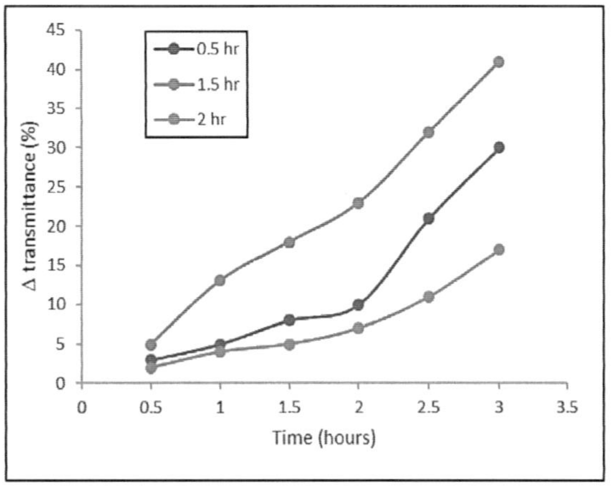

Figure 8. Comparison magnitude transmittance of methyl orange for difference duration of anodized for of Nb_2O_5 nanoparticles.

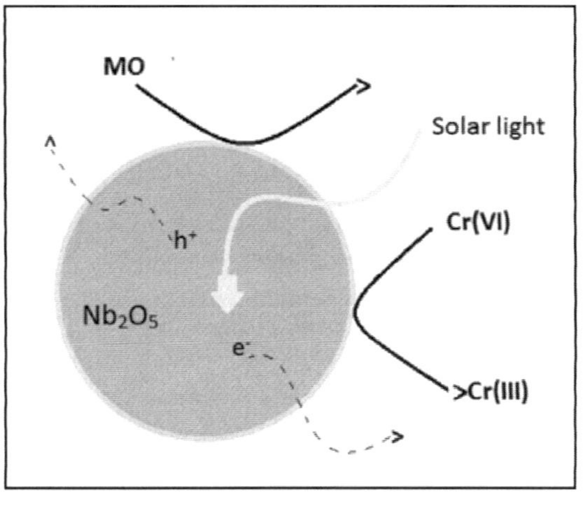

Figure 9. Schematic diagram of charge transfer of photogenerated electrons and holes

3.3. Photocatalytic mechanism.

The photocatalytic activity takes place in solar simulator under solar light. The processes that involve in this activity are reduction and oxidation. In this reaction, the photons have higher or equal energy than the bandgap energy ($hv \geq Eg$).

The mechanism of photocatalytic is shown in Figure 9. The methyl orange solution or dye absorbs the solar light and become excited [15]. As shown in figure 9, the excited dye changes to suitable state whether singlet ($^1Dye^*$) or triplet ($^3Dye^*$) and releases electron [16]. Afterward, the electron is infused onto the conduction band of the semiconductor particles [17]. At the same time, the dye is transformed to the cation dye radicals (Dye^+) that experiences degradation to yield products.

The photocatalyzed reaction is important but may not occur if the recombination of the elctron and hole happen during reaction. When activated electron with an oxidant agent give a product of a reduced product and the generated hole with the reductant agent give the product of an oxidized product, the photocatalyzed will occur

The reaction of photocatalytic can be expressed as shown below (Eqs. (1) – (4)) [18]:

$$Dye + \text{solar light} \rightarrow {}^1Dye^* + {}^3Dye^* \quad (1)$$

$${}^1Dye^* \text{ or } {}^3Dye^* + Nb_2O_5 \rightarrow Dye\bullet^+ + Nb_2O_5(^eCB^-) \quad (2)$$

$$Nb_2O_5(^eCB^-) + O_2 \rightarrow O_2\bullet^- + Nb_2O_5 \quad (3)$$

$$Dye\bullet^+ \rightarrow \text{degradation products} \quad (4)$$

There are significant impacts of nanoporous surface to area ratio toward photocatalytic performance. As discuss above, the nanoporous size represent the absorption surface area. The larger the surface area, the increase negative charge (MO^-) can be absorbed by Nb_2O_5 nanoparticles compare to small surface area of nanoporous. Thus, the degradation of methyl orange solution becomes increase because of this rapid absorption process.

4. Conclusion

The effects of anodized duration toward photocatalytic performance of Nb_2O_5 nanoparticles have been successfully investigated. When the anodized time increase up to 1.5 hours, the size Nb_2O_5 nanoporous also increase. The increase of anodized process caused the performance of photocatalytic increase too. The performance of photocatalytic shown by the transmittance graphs that decrease in value as the anodized duration increase. When the anodized time exceed than 1.5 hours, the anodized process does not affective anymore as the size of Nb_2O_5 nanoparticles become decrease. The performances of photocatalytic also decrease as the transmittance increase. Overall, the increases of anodized duration up to 1.5 hours cause the performance of photocatalytic also increase..

5. Acknowledgement

The authors would like to acknowledge with gratitude to Faculty of Electrical Engineering UiTM for the facilities. This work is fully supported by Ministry of Higher Education Malaysia (MOHE) under the Research Acculturation Grant Scheme (RAGS) (Project code: 600-RMI/RAGS 5/3 (6/2015)).

6. References

[1] "A Composite Photocatalyst for Methylene Blue Degradation Under Visible Light Irradiation", *Asian J. Chem.*, vol. 26, no. 8, 2014.

[2] F. Brites, V. Santana and N. Fernandes-Machado, "Effect of Support on the Photocatalytic Degradation of Textile Effluents Using Nb2O5 and ZnO: Photocatalytic Degradation of Textile Dye", *Topics in Catalysis*, vol. 54, no. 1-4, pp. 264-269, 2011.

[3] H. Chen, L. Qiu, J. Xiao, S. Ye, X. Jiang and Y. Yuan, "Inorganic–organic hybrid NiO–g-C3N4 photocatalyst for efficient methylene blue degradation using visible light", *RSC Advances*, vol. 4, no. 43, p. 22491, 2014.

[4] H. Cui, K. Dwight, S. Soled and A. Wold, "Surface Acidity and Photocatalytic Activity of Nb2O5/TiO2 Photocatalysts", *Journal of Solid State Chemistry*, vol. 115, no. 1, pp. 187-191, 1995.

[5] F. Hashemzadeh, A. Gaffarinejad and R. Rahimi, "Porous p-NiO/n-Nb2O5 nanocomposites prepared by an EISA route with enhanced photocatalytic activity in simultaneous Cr(VI) reduction and methyl orange decolorization under visible light irradiation", *Journal of Hazardous Materials*, vol. 286, pp. 64-74, 2015.

[6] F. Hashemzadeh, R. Rahimi and A. Ghaffarinejad, "Mesoporous nanostructures of Nb2O5 obtained by an EISA route for the treatment of malachite green dye-contaminated aqueous solution under UV and visible light irradiation", *Ceramics International*, vol. 40, no. 7, pp. 9817-9829, 2014.

[7] S. Lam, J. Sin, I. Satoshi, A. Abdullah and A. Mohamed, "Enhanced sunlight photocatalytic performance over Nb2O5/ZnO nanorod composites and the mechanism study", *Applied Catalysis A: General*, vol. 471, pp. 126-135, 2014.

[8] T. Ma, S. Isobe, Y. Wang, N. Hashimoto and S. Ohnuki, "Catalytic Effect of Nb2O5 in MgH2-Nb2O5 Ball-Milled Composites", *Catalysts*, vol. 2, no. 4, pp. 344-351, 2012.

[9] C. Nico, T. Monteiro and M. Graça, "Niobium oxides and niobates physical properties: Review and prospects", *Progress in Materials Science*, vol. 80, pp. 1-37, 2016.

[10] J. Ou, R. Rani, M. Ham, M. Field, Y. Zhang, H. Zheng, P. Reece, S. Zhuiykov, S. Sriram, M. Bhaskaran, R. Kaner and K. Kalantar-zadeh, "Correction to Elevated Temperature Anodized Nb 2 O 5 —A Photoanode Material with Exceptionally Large Photoconversion Efficiencies", *ACS Nano*, vol. 6, no. 6, pp. 5737-5737, 2012.

[11] F. Peng and D. Li, "Synthesis of (Na2O,PbO)–Nb2O5, (Na2O,BaO)–Nb2O5, and (K2O,SrO)–Nb2O5 thin films using sol–gel method", *Thin Solid Films*, vol. 518, no. 23, pp. 6833-6838, 2010.

[12] M. Rahman, R. Rani, A. Sadek, A. Zoolfakar, M. Field, T. Ramireddy, K. Kalantar-zadeh and Y. Chen, "A vein-like nanoporous network of Nb2O5 with a higher lithium intercalation discharge cut-off voltage", *J. Mater. Chem. A*, vol. 1, no. 36, p. 11019, 2013.

[13] R. Rani, A. Zoolfakar, J. Ou, M. Field, M. Austin and K. Kalantar-zadeh, "Nanoporous Nb2O5 hydrogen gas sensor", *Sensors and Actuators B: Chemical*, vol. 176, pp. 149-156, 2013.

[14] C. Shifu and C. Gengyu, "Photocatalytic degradation of organophosphorus pesticides using floating photocatalyst TiO2·SiO2/beads by sunlight", *Solar Energy*, vol. 79, no. 1, pp. 1-9, 2005.

[15] M. Silva, R. Marques, N. Machado and O. Santos, "Evaluation of Nb2O5 and Ag/Nb2O5 in the photocatalytic degradation of dyes from textile industries", *Brazilian Journal of Chemical Engineering*, vol. 19, no. 4, 2002.

[16] S. Suzuki, K. Teshima, K. Yubuta, T. Shishido and S. Oishi, "Fabrication of transparent–colorless Nb2O5 nanocrystal layers and their photocatalytic evaluation using organosilane thin films", *Applied Surface Science*, vol. 280, pp. 539-544, 2013.

[17] H. Wang, P. Xu and T. Wang, "Doping of Nb2O5 in photocatalytic nanocrystalline/nanoporous WO3 films", *Thin Solid Films*, vol. 388, no. 1-2, pp. 68-72, 2001.

[18] J. Xing, Z. Shan, K. Li, J. Bian, X. Lin, W. Wang and F. Huang, "Photocatalytic activity of Nb2O5/SrNb2O6 heterojunction on the degradation of methyl orange", *Journal of Physics and Chemistry of Solids*, vol. 69, no. 1, pp. 23-28, 2008.

IOP Conf. Series: Materials Science and Engineering **340** (2018) 012008 doi:10.1088/1757-899X/340/1/012008

Investigation of Annealing Temperature on Copper Oxide Thin Films Using Sol-Gel Spin Coating Technique

H. Hashim, S. F. A. Samat, S. S. Shariffudin and P. S. M. Saad

Faculty of Electrical Engineering, Universiti Teknologi Mara,
40450 Shah Alam, Selangor, MALAYSIA

E-mail: hashimah655@salam.uitm.edu.my

Abstract. Copper (II) Oxide or cupric oxide (CuO) is one of the well-known materials studied for thin films applications. This paper was studied on the effect of annealing temperature to CuO thin films using sol-gel method and spin coating technique. The solution was prepared by sol-gel method and the thin films were synthesized at various temperatures from 500°C to 700°C that deposited onto the quartz substrates. After the annealing process, the thin films were uniform and brownish black in colour. The measurements were performed by atomic force microscopy (AFM), surface profiler (SP), two-point probe and Ultraviolet-visible (UV-Vis-NIR) spectrometer. From the optical measurement, the band gap was estimated to be 1.44eV for sample annealed at 550°C.

1. Introduction

CuO thin films are semiconductor material which normally used as an active layer in solar cells applications [1]. The investigation of the annealing temperatures on CuO thin films were usually deposited by many different methods such as dip coating, spin coating, chemical vapor deposition, electrochemical deposition, sputtering, and spray pyrolysis [5]. Spin coating technique was used in this paper fabricated with sol-gel method. There were many advantages by using sol-gel technique such as stable to decompose by sunlight, stable to heat, low temperature process and highly transparent [5]. Based on the research by Mohamammad Hossien Habibi et al [2] state that the copper oxide compounds could be used in the high temperature superconductors.

CuO thin films found to be a native p-type conductivity due to copper vacancies in the structure [11-14] with a band gap between 1.3eV and 2.1eV [3]. The optical absorption of CuO nanoparticles were controlled by the size and shape of nanostructure and the different in annealing temperature determined its activity [2]. Several samples of CuO thin films with temperatures range of 500°C to 700°C were reported in this paper. The research was to improve the microstructure of the CuO with annealing temperature.

The characterization of the CuO thin films were measured by I-V measurement, optical properties, surface morphology and thickness of thin films. The CuO thin films were characterized by its electrical and surface morphology using two-point probe technique and AFM respectively. Meanwhile, the thickness of the CuO thin films conducted by using SP. Lastly, the optical band gap energy was calculated by transmittance that conducted by UV-VIS NIR spectrometer.

IOP Conf. Series: Materials Science and Engineering **340** (2018) 012008 doi:10.1088/1757-899X/340/1/012008

2. Experimental Details

The procedures for preparing copper oxide thin films were summarized in this section.

2.1. Solution preparation

The quartz substrates with 2cm x 2cm dimension started with cleaning process. The five samples were immersed in acetone, methanol and distilled water sonicated with ultrasonic cleaner consecutively. Then, the solution was prepared by dissolving Monoethanolamine (MEA, C2H7NO) and Isopropanol in Copper Acetate Cu(CH3COO)2. H2O. Afterwards, (poly)ethylene glycol (PEG, H(OCH2CH2)nOH) was added to the prepared solution. Then it was mixed for 10 minutes using a magnetic bar and filtered on the hotplate with stirrer, Stuart (UC 152). The solution was prepared for 30 to 35 minutes long since no refluxing was required. Table 1 shows the ratio for chemical used in solution. Meanwhile, the powder of Copper Acetate was 0.25 gram. MEA was observed to be very effective and resulted in a very fast dissolution of copper acetate. The prepared solution with concentration of 0.25M was dark blue and clear. A double-step spinning program was applied to obtain homogeneous precursor films prior to heat treatment.

Table 1. The chemical ratio and quantity used in the CuO solution.

Chemical used	Ratio and quantity of chemicals	
	Percentage (%)	Mass (ml)
Isopropanol	5	4.5
Monoethanolamine (MEA, C_2H_7NO)	90	0.25
(poly)ethylene glycol PEG	5	0.25

2.2. Deposition of thin films

In first step, the solution was dropped in 10 times onto the substrates at 100 rpm for 10 seconds. The coating process of the thin films was used spin coater (WS-650-23B). Then, coating and drying of the precursor films were achieved during the second spinning step at 3000 rpm for 300 seconds on microscope quartz slides. Further drying and decomposition reactions occurred during pyrolysis in a chamber furnace, PROTHERM (PLF 160/25) at 250 °C for five minutes following deposition each layer. After coating of the five layers which were the last layer of each sample, the films were annealed from 500°C to 700°C for 30 minutes.

2.3. The characterization of CuO thin films

The thicknesses of the thin films were measured using surface profiler (Dektak 150). Surface morphologies of the thin films were investigated using AFM. An UV-VIS-NIR Spectrophotometer (JASCO, FLH-740) was used to record the transmittance and absorbance between wavelength of 200nm and 1000nm. The PL spectra (CDRH Microscope Enclosure) were characterized in the wavelength range of 326nm to 900nm at room temperature. The current-voltage (I-V) measurement was conducted by using two point probes. Figure 1 shows the side view of the sample with Platinum as the metal contact for I-V measurement.

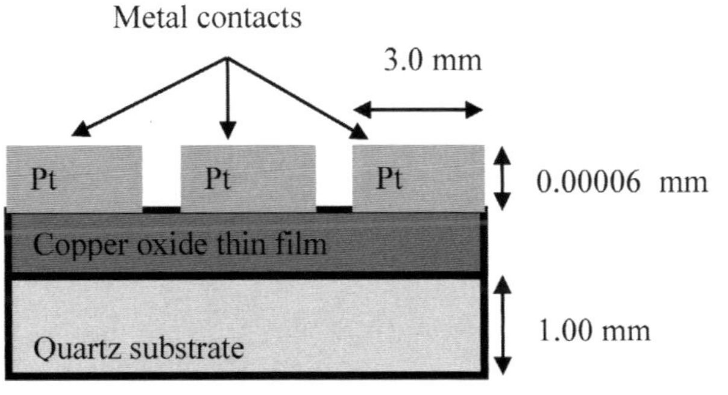

Figure 1. The side view of CuO thin film.

3. Results and Discussion

The results obtained from the experiment are discussed in this section.

3.1. Thickness measurement

Figure 2 shows the thickness of CuO thin films annealed at different temperatures. The graph shows that the thickness of CuO will increase as the temperatures getting higher. The thickness measurement was one of the parameter that could change the properties of the material due to surface phenomena [10]. The thickness of the nanocrystalline CuO thin films were measured using SP and found to be 70.7nm, 135.5nm, 173.2nm, 199.1nm and 255.3nm for 500°C, 550°C, 600°C, 650°C and 700°C respectively.

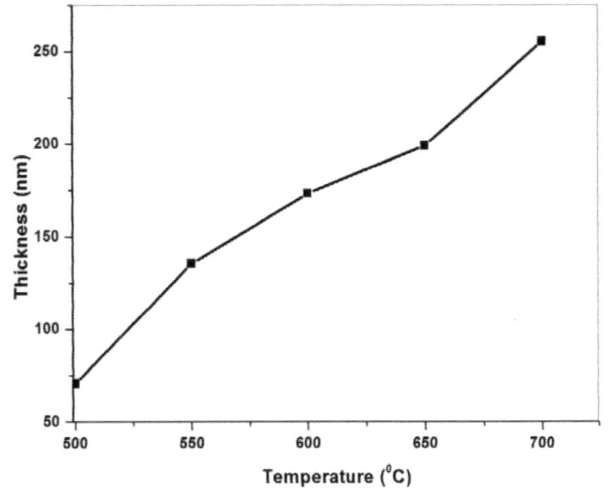

Figure 2. The thickness of CuO thin films at different annealing temperature.

3.2. Morphological Studies

AFM measurements were used to analyze surface morphology of the film films. Figure 3 shows the results of three dimensional (3D) AFM image of CuO film for different annealed temperatures. The surface evolution of the films showed "hills and valley" like structures, which are uniformly distributed over the entire substrate surface.

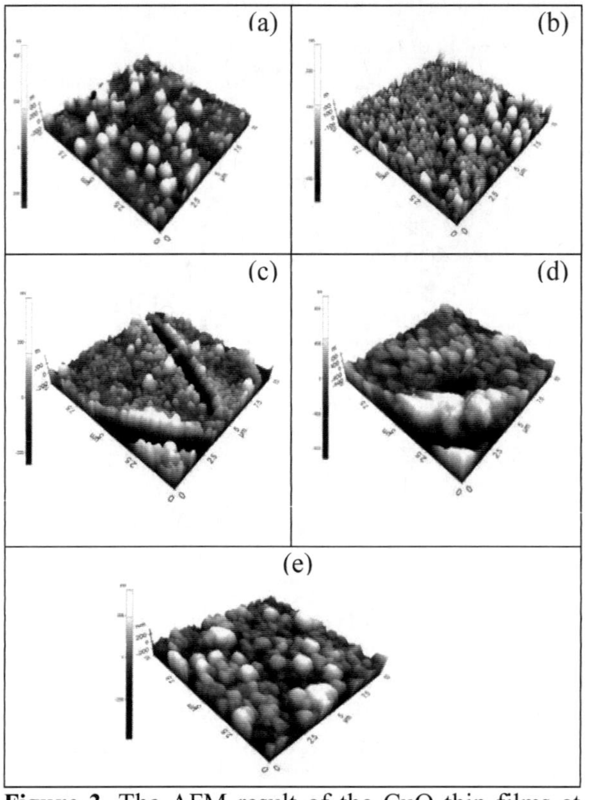

Figure 3. The AFM result of the CuO thin films at different temperature (a) 500°C, (b) 550°C, (c) 600°C, (d) 650°C and (e) 700°C.

The grain size of CuO microstructures were getting bigger and harden with increasing of annealing temperature. Table 2 shows the mean value of thickness and surface roughness with the different temperatures. The thicknesses increases, same goes to the surface roughness also increases with increase in grain size. The sample A, B, C, D and E represent different temperatures of 500°C, 550°C, 600°C, 650°C and 700°C respectively.

Table 2. The data of thickness and surface roughness.

Sample	Annealed temperature (°C)	Thickness (nm)	Ra (nm)
A	500	70.7	42.63
B	550	135.5	52.27
C	600	173.2	60.43
D	650	199.1	67.34
E	700	255.3	76.52

Ripening of the grains causes growth from smaller to larger grain by surface diffusion which reduces the surface to volume ratio. Also higher substrate temperature stimulates elimination of grain boundaries and result in the formation of larger grains. These observations are in agreement with Chacko et al [12]. However, the results show that the thin films were uniform, dense, and well adhered to the quartz substrate.

IOP Conf. Series: Materials Science and Engineering **340** (2018) 012008 doi:10.1088/1757-899X/340/1/012008

3.3. Optical properties

Figure 4 shows the value of optical transmittance increases with higher the annealing temperatures. The CuO thin films prepared at a wavelength ranging from 300nm to 800nm, due to their reduced surface area.

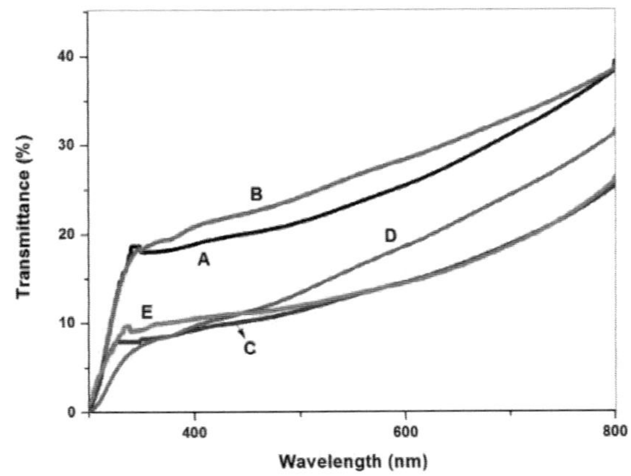

Figure 4. The transmittance of CuO thin films deposited at different annealed temperature (A) 500°C, (B) 550°C, (C) 600°C, (D) 650°C and (E) 700°C.

Figure 5 illustrates the absorption coefficient (α) was calculated from the optical transmittance spectra by using UV-Vis Spectrometer. The absorption values were decreases in the visible region with increasing annealing temperatures. The results show that all samples have an abrupt absorption edge in the range of 400nm to 900nm. The results show the best plot temperature annealed at 500°C and 550°C. Meanwhile, the annealed temperature at 600°C, 650°C and 700°C showed different trend from of the thickness. This showed the microstructures of CuO affect with the high temperature. From both Figure 4 and 5 show the optical transmittance increasing but the absorption coefficient decreasing with the higher annealing temperature.

As a direct band gap semiconductor, the relationship between absorption coefficient of the CuO thin films and incident phonon energy (hv) can be express as follow Equation 1 [5]:

$$hv = \frac{hc}{\lambda} \tag{1}$$

where h represent the Plank constant which was 6.63×10^{-34}, c was the light constant 3×10^{8} and λ was the wavelength. Meanwhile the absorption coefficient, α was calculated by using Lambert`s law as shown by Equation 2 below:

$$\alpha = \frac{1}{t} \ln \left(\frac{1}{T}\right) \tag{2}$$

where t was the thickness of thin film and T was the transmittance values of the thin films. The Urbach energy is a parameter used to describe the band tailing phenomenon which normally defects in the thin films [10].

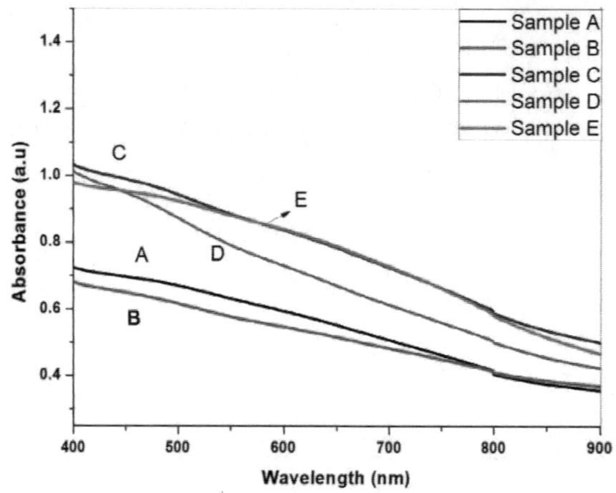

Figure 5. The absorbance deposited of CuO thin films measured by UV-Vis NIR Spectrometer at different temperature..

Figure 6 shows the reciprocal gradient of the plot in Figure 5 with the value of Urbach energy or optical energy band gap. From the plots, the optical band gap energy value of as deposited samples annealed at 500°C, 550°C, 600°C, 650°C and 700°C are 1.39eV, 1.44eV, 1.70eV, 1.85eVand 1.90eV respectively.

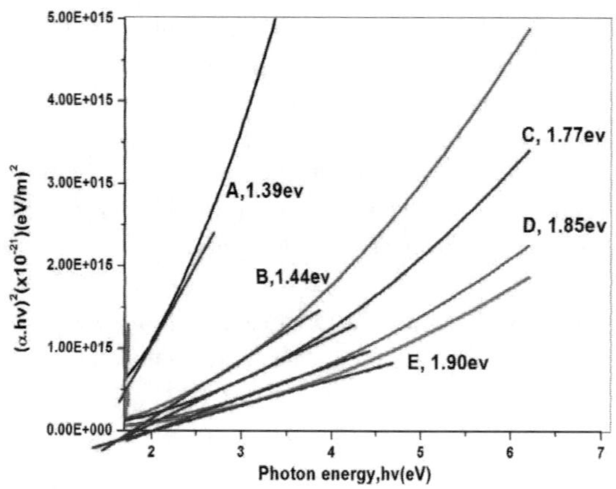

Figure 6. The plot of the (αhv) versus hv for CuO thin films.

The results show the optical energy band gap increases with annealing temperatures. The band tail widening from the plot shows that the defect of Cu (CH3COO)2. H2O doped CuO thin films increase with annealing temperatures. Based on the past research, this phenomenon could be from the broadening of donor levels into impurity band which merge with conduction band in the forbidden gap [5]. More Cu+ could be incorporated in CuO lattice structure by higher thermal energy obtained at higher annealing temperature due the increment of carrier concentration.

IOP Conf. Series: Materials Science and Engineering **340** (2018) 012008 doi:10.1088/1757-899X/340/1/012008

3.4. Electrical properties

I-V measurement plots of CuO thin films at different annealing temperature are shown in Figure 7. Platinum (Pt) was used as metal contacts during the measurement. The result indicates the Ohmic behavior of thin films with linear I-V curve observed for all annealed thin films. It also could be observed that the current value at fixed voltage increase with annealing temperatures to indicate the enhancement of electrons conductivity as annealing temperatures were increased. From the I-V curve, the resistivity, ρ of CuO thin films were calculated using following Equation 3:

$$\rho = \left(\frac{V}{I}\right)\frac{wt}{l} \tag{3}$$

where ρ was the resistivity, V was supplied voltage, I was measured current, t was the film's thickness, w was the electrode width and l was the length between electrodes.

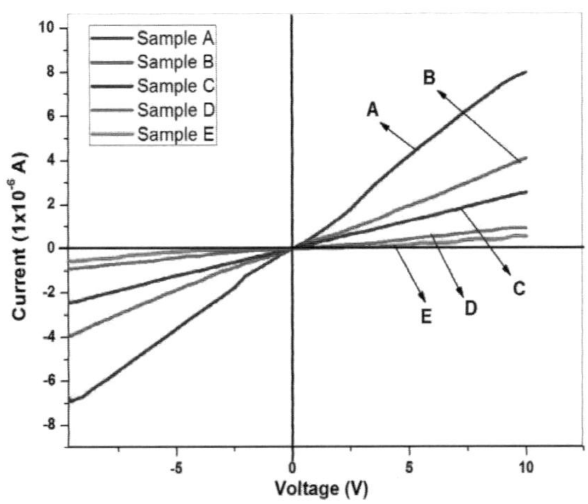

Figure 7. The I-V curve of CuO thin films at different annealing temperature.

The calculated resistivity obtained in this worked to be 2.27kΩm, 5.56kΩm, 7.86kΩm, 14.4kΩm and 38.3kΩm for thin films annealed at 500°C, 550°C, 600°C, 650°C and 700°C respectively. The plot of CuO thin films resistivity with annealing temperature was shown in Figure 8. The temperature at 700°C has the highest resistivity while the temperature at 500°C has the lowest resistivity. It could be concluded that the resistivity of the CuO thin films are increase with increase of annealing temperatures.

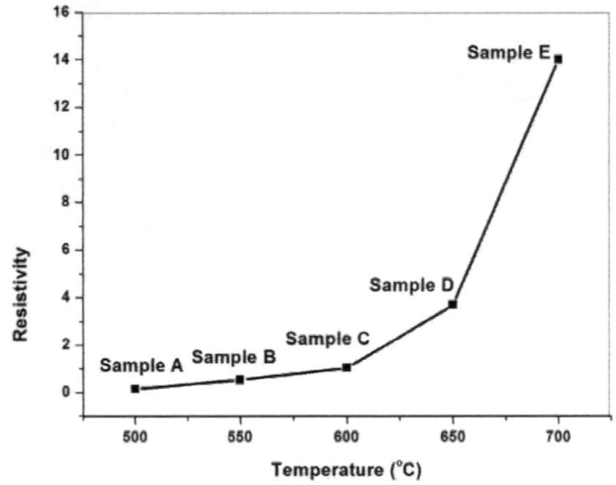

Figure 8. The resistivity of CuO thin films as an increasing of annealing temperatures.

The decreased in electrical conductivity at higher annealing temperatures could be due to the crystallinity with annealing temperature [5]. Dandeneau et al. [15] shows that the increment of carrier concentration by the ionization of oxygen vacancies with annealing temperature also contributed to the increment of electrical conductivity at higher annealing temperature.

4. Conclusion

The thin films were annealed at 500°C until 700°C. The thicknesses were increased by increasing temperature. The AFM revealed that the CuO thin film has a good surface morphology. All the thin films exhibit high transparency in the visible region with average transmittance above 16%. The energy optical band gap of CuO thin films increased as annealing temperature getting higher such as 1.44eV for sample annealed at 550°C. It was proved by current-voltage (I-V) measurement indicates conductivity of CuO thin films has the highest plot at 500°C and decreasing with the increasing annealing temperatures.

Acknowledgments

The authors would like to thank to the NANO-Electronic Centre (NET), Faculty of Electrical Engineering, Universiti Teknologi MARA and NANO-SciTech Centre, Institute of Science, Universiti Teknologi MARA for providing the laboratory facilities. This research was funded by the Lestari Grant (File No.: 600-RMI/DANA 5/3/LESTARI (26/2015)) under UiTM internal grant scheme.

References

[1] Karthibk Kumar S, Suresh S, Murugesan S and Samuel Paul Raj 2013 *Sciverse Science Direct Solar energy* **94** c299-304
[2] Mohammad Hossien Habibi and Bahareh Karimi 2013.
[3] Oral A Y, Mensur E, Aslan M H, Basaran E 2013.
[4] Lidia Armelao, Davide Barreca, Manuel Bertapelle, Gregorio Bottaro, Cinzia Sada and Eugenio Tondello 2003 *Thin Solid Films* **442** 48-52.
[5] Mamat M H, Sahdan M Z, Amizam S, Rafaie H A, Khusaimi Z, Ahmed A Z, Abdullah S, and Rusop M 2008 *IEEE International Conference on Semiconductor Electronics* 566–570
[6] Guille'n C. and Herrero 2008 *Vacuum* **82** 668
[7] Hiroyuki Iechi, Masatoshi Sakai, Kenji Nakamura, Masaaki Iizuka, Masakazu Nakamura and Kazuhiro Kudo 2005 *Synthetic Metals* **154** 149

[8] Keis K, Bauer C, Boschloo G, Hagfeldt A, Westermark K, Rensmo H and Siegbahn H 2002 *Journal of Photochemistry and Photobiology A: Chemistry* **1** 48 57
[9] Foo K L, Kashif M, Hashim U, Usman Ali S M and Willander M 2012 *International Conference on Biomedical Engineering* 223–226
[10] Md Sin N D, Mamat M H and Mohammad Rusop 2001 *Advanced Materials Research* **667** 3 507–510
[11] Iqbal Singh and Bedi R K 2011 *Department of physics, Khalsa College, Amritsor* 143005 Punjab India
[12] Science M 2007 *poland* **25** 3 2–7
[13] Attia J M S M, Jue Wang, Guagming Wu and Jun Shean 2002 **18**
[14] Bravina S L, Morozovsky N V, Eliseev E A, Morozovska A N, Costecalde J, Soyer C, Remiens D, Deresmes D 2012 *Journal of Applied Physics* **112** 5 052015 052015-7
[15] Dandeneau C S, Jeon Y H, Shelton C T, Plant T K, Cann D P, & Gibbons B J 2013 *Thin Solid Films* **517** (15) 4448-4454
[16] Halin D S C, Talib I A, Daud A R, & Hamid M A A 2013 *Key Engineering Materials* 594-595 113-117
[17] Sawsan D, Yousef H, Ahmad I A, Nacir T 2014 149-154

IOP Conf. Series: Materials Science and Engineering **340** (2018) 012009 doi:10.1088/1757-899X/340/1/012009

Crack-Free TiO_2 Thin Film via Sol-Gel Dip Coating Method: Investigation on Molarity Effect

M.S.P. Sarah, B.A.M. Hasmi, S.S. Shariffudin, H. Hashim, S.H. Herman

NANO-ElecTronic Centre, Faculty of Electrical Engineering, Universiti Teknologi MARA, 40450 Shah Alam, Selangor, Malaysia

puterisarah@salam.uitm.edu.my

Abstract. Titanium dioxide, (TiO_2) is transparent to visible light and excellent in optical transmittance. In this experiment, TiO_2 thin films have been fabricated on glass substrates by sol-gel dip coating method. Sol-gel dip coating technique can provide corrosion protection performance and can produce thick coating. Due to the thick coating, the possibility to get a crack thin film is high. Therefore, this paper is to investigate a crack-free TiO_2 thin film for optical sensor application where it is needed the most. The films were subjected to different molar concentration solution with 0.01M, 0.05M, 0.10M, 0.15M and 0.20M. The thin films are then characterized by field emission scanning electron microscopy (FESEM), atomic force microscopy (AFM), surface profiler and UV- VIS NIR spectrometer. From this study, a crack-free of TiO_2 thin film was obtained and this thin film could be used for optical sensor application later.

1. Introduction

TiO_2 is a wide band-gap semiconductor that is of interest for different kinds of optical applications. Some of the examples are photocatalysts, optical spacers in polymer photovoltaic cells, and optical sensors [1]. TiO_2 is also an important commercial element because of its characteristics which are highly transparent and high refractive index (n=2.6142) [2]. Example of TiO_2 applications are pigment in paints and coating materials in optical thin films. The distinct properties of TiO_2 as its high refractive index and photocatalytical activity lead to high types of applications in various areas such as electronics sensing optics and catalysis [3]. Due to excellent optical transmittance in the visible range [4], transparency over a wide spectral range and high reflective index and the absorption of UV light [5], TiO_2 films become essential in optical films. The distinct properties of TiO_2 make a particular sensing purpose can be reached with the types of TiO_2 phase [6], composition and nanostructure feature are decisive factors for sensor performance [7].

In previous work, it was reported that the preparation and characterization of TiO_2 films have been largely used due to increasing number of applications in optics [8]. Chemical vapour deposition, sputtering and sol-gel is some of the method to prepare TiO_2 materials [9]. A flexible method like sol-gel method can control the uniformity and thickness of the thin film, and thermal stability of high surface area material [10]. Sol-gel method has many benefits such as very simple in operation [11], can be used for deposition of substrates, and easily anchored on the substrate [12]. On top of that, this method enables the modification of the structure and the microstructure of the deposited TiO_2. The properties of the sol-gel TiO_2 thin are highly dependent on the structure, thickness and the density of the deposited layers [13]. These features are predominantly manipulated by the sol composition, the sol viscosity and the withdrawal speed in the dip coating process [14].

Researchers have studied that controlling of the viscosity is very important in the deposition process. The controlling of parameter includes characteristics of the solution such as ionic concentration, conditions of heating and withdrawal speed, where this is compulsory in order to acquire crack-free films [15]. From observation, the thin film thickness is directly proportional to the viscosity of the precursor sol [16]. Moreover, the appearance of cracks occurred when the critical value for thickness of the thin films is reached [17]. In fact, withdrawal speed has to be controlled to prevent the thickness of the films producing cracks and bubbles after annealing [18]. Furthermore, by using sol-gel dip coating technique, the microstructure of the film can be easily controlled by changing the solution composition and deposition condition. These parameters also were influenced to obtain a crack-free TiO_2 thin film [19]. There are a few types of crack that have been recognized. Rosette is one type of crack where it is short and branching from the center [20]. This type can be observed over the whole surface of the thin film and occurs at lower viscosities. The other type of crack is a long straight line and these cracks obtained from solution with higher viscosity [21].

This research is to investigate the crack- free thin film by sol-gel dip coating technique with different molar concentration. After optimizing the characteristics of TiO_2 thin films based on optical properties, surface morphology, topographic surface and thickness of the film, the TiO_2 thin film will be used for optical sensor application.

2. Methodology

2.1. Glass Substrate Preparation
The glass was cut with the size of 2 cm x 2 cm. Then, it was cleaned with acetone in ultrasonic bath, followed by ethanol and lastly with di-ionized (DI) water for 10 minutes. Lastly, it was blew with N_2 gas for drying purpose.

2.2 TiO_2 Solution by Sol-Gel Method Preparation
The starting material used was Titanium (IV) Isopropoxide (TTIP). The molarity of the precursor was varied from 0.01M, 0.05M, 0.1M, 0.15M and 0.2 M. Glacial acetic acid and Triton-X were added to the solution as stabilizer and surfactant respectively. Small amounts of DI water were also added to increase the hydrolysis reaction rate.

2.3 Deposition of TiO_2
TiO_2 thin films were deposited on a glass substrate by pulling it from the solution at constant withdrawal speed and dip speed of 10 mm/s by using the dip coater equipment, Precision Dip Coater. The dipping cycles was set at 15 cycles. The glass was attached at the holder perfectly to prevent it from falling in the solution. The thin films were dried at 100° C for 30 minutes in a chamber furnace, PROTHERM (PLF 160/25). It was then undergoes the annealing process at 450° C for 2 hours to restructure the crystallinity of the thin film.

2.4 Sample Characterization
The samples were characterized using atomic force microscope (AFM, Park System XE-100) for structural images. Meanwhile, the thicknesses have been measured using Dektak 150 Stylus Veeco Dektak 750 Surface Profiler. As for the surface morphology, FESEM was done using JEOL JSM-7600F. The optical properties were studied using UV-Visible (Vis)-near infrared.

3. Results and discussions

A. FESEM

Figure 1 shows the FESEM micrographs of the surface morphologies of the TiO_2 thin film for each molarity. It can be observed from the FESEM images that the films are free from the crack. The grain size of all the samples were also measured.

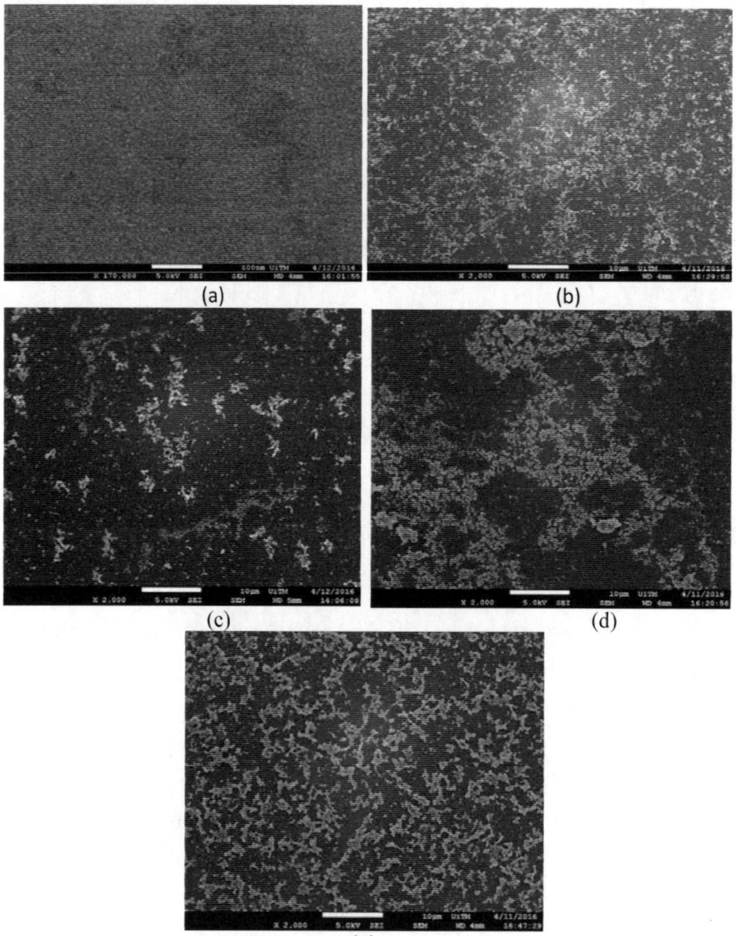

Figure 1. FESEM image of TiO_2 thin film at different molarity (a) 0.01M, (b) 0.05M, (c) 0.10M, (d) 0.15M, and (e) 0.20M

Figure 1(a) shows that there was no surface to show crack-free of TiO_2 thin film. This was due to the lower molarity of solution. When the solution was prepared, the viscosity of the solution is too low. This can be seen where the glass substrate shows very transparent features after completing the dip coating process. The glass substrate was very transparent and looks as the same as glass before the glass was coated. Even though, 6nm of platinum was coated on the glass and the image was zoom in, there is still no structure on the image view. However, the image started to show a crack-free surface of TiO_2 thin film at 0.05M, 0.10M, 0.15M and 0.20M, Figure (b), Figure (c), Figure (d) and Figure (e) respectively.

As seen in Table 1, the distribution of the grain size of the thin film is depending on molarity. It was observed that when molarity is increasing, the grain size also increasing. By analyzing and characterizing the grain size of each molarity, it can be seen that the grain size of the lowest molarities of 0.01M cannot be determined. This is because the size of the grain for 0.01M is too small due to TiO_2

thin film is very transparent. However, when the molarity is 0.05M, it tends to agglomerate and the grain size is 116nm which is the smallest grain size. The agglomeration is due to mixing technique by using the magnetic stirrer that the solution was not stirred continuously that makes nanoparticles easily agglomerates [22]. Besides that, the agglomeration is significantly influenced by film thickness and agglomeration easier occurs when the film thickness is thinner [23]. When the molarity is 0.10M, the grain size was increased by 137 nm. Moreover, when molarities are 0.15M and 0.20M, the grain size are 145nm and 152nm respectively which gives grain size bigger than the other molarity. For 0.15M, it can be seen that the grain growth more on the surface of thin film and quite uniform than the other molarity. It was observed that higher the molarity of TiO_2, the uniformity of the surface morphology is better.

TABLE 1. THE DATA OF GRAIN SIZE OF THE THIN FILM DEPEND ON MOLARITY

Sample	Molarity (M)	Average Grain Size (nm)
A	0.01	Too small
B	0.05	116
C	0.10	137
D	0.15	145
E	0.20	152

B. Topographic Surface

Figure 2: AFM image for 0.15M sample

Figure 2 shows the AFM topographic images of TiO_2 thin film after annealing at 450° C temperature for 0.15M sample. The scanning area is 10um x 10um. The surface roughness was measured to study the morphology evolution. The value of the surface roughness at this molarity is 157.094 nm and thickness is 802.01 nm. Due to 0.15M is much more uniform than the other sample, so the roughness of this sample was measured. In the previous research, it was reported that when the molarity of the solution is increasing, the roughness of the thin film is decreasing [24]. By referring to Figure 2; the film is quite uniform with not much porosity, so that the roughness was low.

A. Optical properties

UV- Vis Analysis

Figure 3 shows the transmittance of each sample in the wavelength range from 300 to 800 nm. The transmittance value is taken at wavelength around 350 nm. At 0.01M, the optical transmittance value is almost 100.0% which gives the highest reading than the other molarity. The glass substrate is too transparent after the dip-coating process and look like a clean glass. When the sample was characterized by using UV-Vis spectrometer, the transmittance value was higher because the coating of TiO_2 solution is so thin and therefore it cannot be determined by the laser. The transparency starts to decrease for 0.05 M with 60 T%. However, the transmittance values do not show any trend of decreasing when it increase to 75 T% and 85 T% for 0.1 M and 0.15 M respectively. While, 0.20M shows transmittance value 65 T%.

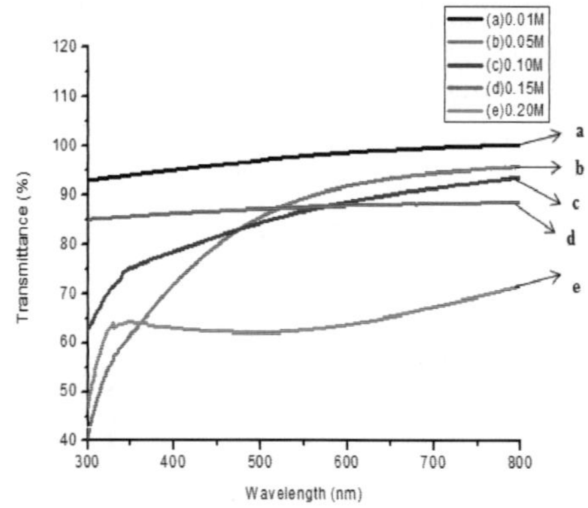

Figure 3: Transmittance spectra of the TiO_2 thin films at different molarity

Figure 4 represents the absorbance spectra of each sample measured by using UV-VIS spectrometer. It can be seen that all the samples have a different absorption edge in the range wavelength of 300 nm to 800 nm. The lowest absorbance belongs to 0.01 M samples with almost 0.00 a.u at 350 nm. Meanwhile, the absorbance value for 0.05M is 0.23 a.u. At 0.1M and 0.15M, the absorbance value is decreasing to 0.13 a.u and 0.03 a.u respectively. But, at 0.2M the absorbance value tends to show increment with the reading 0.2 a.u. The results show no trend for both transmittance and absorbance spectra due to the not well prepared samples. However, the results follows the theory where, the higher the optical transmittance value; the lower absorbance value [25].

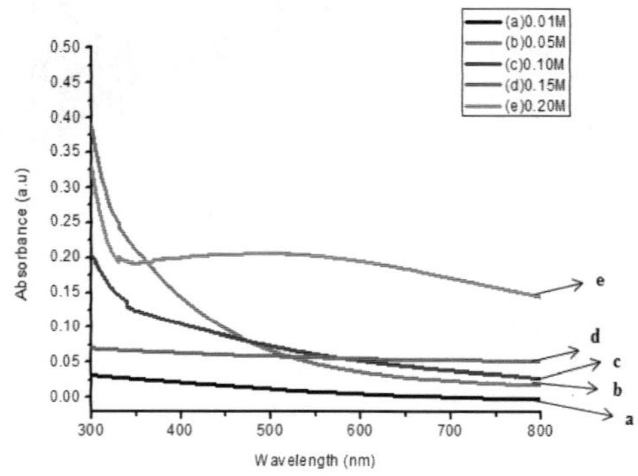

Figure 4: Absorbance spectra of the TiO_2 thin films at different molarity

4. Conclusion

The fabrication of TiO_2 thin film was successfully prepared on glass substrate by using sol-gel dip coating technique. It can be observed that sol- gel dip coating technique was a suitable method to produce a crack-free of TiO_2 thin films. For FESEM images, the higher the molarity of solution, the uniformity of the surface morphology is better. It shows that 0.15M is quite uniform and the grain growth more on the surface of the thin film. Moreover, based on the FESEM result, 0.15M shows low and quite uniform with not much porosity sample. For UV-Vis result, it can be prove that this work follows the theory; high transmittance, low absorbance.

Acknowledgement
The authors would like express their deepest appreciation to all the staffs of NANO-ElecTronic Centre (NET) and NANO-SciTech Centre (NST) for their assistance in completing this work. Special thanks also go to the Faculty of Electrical Engineering, UiTM Shah Alam and grant 600-RMI-FRGS 5/3 (0106/2016) for the funding.

References

[1] E. Haimi, H. Lipsonen, J. Larismaa, M. Kapulainen, J. Krzak-ros, and S. Hannula, *Thin Solid Films*, vol. 519, no. 18, pp. 5882–5886, 2011.

[2] K. K. Saini, S. D. Sharma, M. Kar, D. Singh, and C. P. Sharma, vol. 353, pp. 2469–2473, 2007.

[3] L. R. Paez and J. Matoušek, *Ceramics*, vol. 48, no. 2, pp. 66–71, 2004.

[4] S. Vyas, R. Tiwary, K. Shubham, and P. Chakrabarti, *Superlattices Microstruct.*, vol. 80, pp. 215–221, 2015.

[5] R. Mechiakh, N. Ben Sedrine, R. Chtourou, and R. Bensaha, *Appl. Surf. Sci.*, vol. 257, no. 3, pp. 670–676, 2010.

[6] S. Islam and R. Abdul, pp. 162–168, 2013.

[7] R. S. Sonawane, S. G. Hegde, and M. K. Dongarevol. 77, pp. 744–750, 2002.

[8] A. Shokuhfar, M. Alzamani, E. Eghdam, M. Karimi, and S. Mastali, vol. 2, no. 1, pp. 16–21, 2012.

[9] A. Z. Simões and J. A. Varela, vol. 2219, no. October 2015, 2001.

[10] M. Shinen, K. Ajeel, and F. Rasin, *Uobabylon.Edu.Iq*, no. Rancourt 1987, pp. 1–9, 2007.

[11] M. Yu, M. Evtushen, S. V Romashkin, and E. Instit, vol. 73, pp. 100–107, 2015.

[12] Yimit, Rossberg, T. Amemiya, and K. Itoh, *Talanta*, vol. 65, no. 5, pp. 1102–1109, 2005.

[13] I. H. H. Affendi, M. S. P. Sarah, N. E. A. Azhar, I. Saurdi, A. Ishak, and M. Rusopno. Istmet, pp. 183–187, 2014.

[14] A. Welte, C. Waldauf, C. Brabec, and P. J. Wellmann, vol. 516, pp. 7256–7259, 2008.

[15] W. C. Pramod, K. Charles, C. Sorrell, and C. C. Sorrell, pp. 2465–2480, 2016.

[16] S. Technique, "Sol-Gel Deposition," pp. 2–5, 2006.

[17] M. Hemissi, H. Amardjia-adnani, and J. C. Plenet, vol. 9, pp. 717–721, 2009.

[18] H. Search, C. Journals, A. Contact, M. Iopscience, I. O. P. Conf, and I. P. Address, vol. 012009.

[19] O. Waveguides, vol. 3002, pp. 2998–3003, 2005.

[20] D. Constant, 2016.

[21] J. Bai and B. Zhou, *Chem. Rev.*, vol. 114, no. 19, pp. 10131–76, 2014.

[22] S. Latthe, S. Liu, C. Terashima, K. Nakata, and A. Fujishima, *Coatings*, vol. 4, no. 3, pp. 497–507, 2014.

[23] Z. Simo*J. Eur. Ceram. Soc.*, vol. 21, no. October 2015, pp. 1151–1157, 2001.

[24] V. Semiconductors and D. P. Values, "Vishay Semiconductors Optical Sensors - Reflective," *Current*, pp. 48–54.

[25] K. Thamaphat, P. Limsuwan, and B. Ngotawornchai, vol. 361, pp. 357–361, 2008.

IOP Conf. Series: Materials Science and Engineering **340** (2018) 012010 doi:10.1088/1757-899X/340/1/012010

Fabrication of MEMS cantilever using laser micromachine

S. Johari, M. Z. Zainol, A. A. Azman, M. Mazalan and Y. Wahab

School of Microelectronic Engineering, Universiti Malaysia Perlis, Malaysia

shazlinajohari@unimap.edu.my

Abstract. This paper presents the fabrication of MEMS cantilever using laser micromachine. This technique of micromachining is able to overcome the problem limitations of conventional lithography. It also facilitates three-dimensional (3D) microfabrication rather than two dimensional (2D) microfabrication of conventional lithography. Prior to fabrication process, wet etching process using KOH solution are carried out on silicon wafer. Etching process is necessary to thin the silicon wafer for the laser micromachine purpose. The etch performance on silicon wafer was investigated by varying the concentration of potassium hydroxide (KOH) solution with respect to time. It can be seen that with higher KOH concentration and higher KOH solution temperature, the etch rate is higher and it will thin the silicon wafer faster. Even though it is beneficial when the time taken for the etching process is faster, this also resulted in a rougher wafer surface. The optimized etch rate is approximately 1µm/min which yield in low surface roughness. The optimized parameters of laser micromachining were implemented to produce MEMS cantilever. Silicon wafer is used because most of the MEMS devices are silicon-based substrate. Three types of microcantilever were fabricated using laser micromachine namely rectangular cantilever, T-shaped cantilever and triangular microcantilever. Scanning electron microscope (SEM) and high power microscope (HPM) were used to obtain the surface morphology on the ablated area of the microcantilevers.

1. Introduction

Device miniaturization is imperative towards the production of micro-electro-mechanical system devices. The existing fabrication process encompass series of manufacturing process such as deposition, etching, mask patterning, then followed by etching and post baking. This is time consuming and costly as it requires various chemicals. For example, for a certain device fabrication process, several layers of photoresist, followed by oxide layer and metal layer have to be deposited on a bare substrate. In addition, the conventional fabrication technique is restricted to only one material surface for bulk etching in three silicon planes directions.

Laser micromachining process has been widely adapted to develop microstructures where it utilizes laser ablation as the fundamental fabrication step [1-5]. The process of laser micromachining encompass the interaction between laser and material, where when the targeted material is exposed to laser radiation, the substrate/material will absorb the energy from the laser. The absorbed laser energy is transformed into lattice vibration energy (thermal). This will result in rising of the temperature, hence heating the targeted surface. The repetition of this process will ultimately melt and vaporize the targeted material (Figure 1). The interactions of laser and its substrate depends on both the substrate material and the laser parameters, such as the laser wavelength, the pulse duration and laser intensity. The process of laser micromachine has been established for various types of materials. Patterning using laser allows direct prototyping of materials without the use of additional mask, photoresist molds and further post-processing treatment.

Fabrication process using laser micromachine will give complex three-dimensional structure without a mask. In addition, laser micromachining will give good quality in high resolution, high precision, high process speeds and less fabrication processing steps. Silicon wafers are favourites materials for MEMS devices since it is easily available and compatible with the processing technology in the microelectronic industries. Moreover, silicon is widely used since it has good electrical conductivity and very wide functional range in the semiconductor dopants

A microcantilever is a device that can be used for physical, chemical or biological sensor by detecting the changes in cantilever bending or vibrational frequency. In terms of sensitivity, microcantilever-based sensing devices has good sensitivity as opposed to existing sensors as their response time is shorter and they required minimal operational power. However, the sensitivity can still be improved by modifying the shapes of the microcantilever. The sensitivity depends on the microcantilever stiffness, and the stiffness can be affected by the geometrical structure of the cantilever, namely its width, length and thickness.

In this paper, we aim to use laser micromachine as an alternative method for the development of arrayed microcantilever towards biomedical application. Previously we have reported on the fabrication of microfluidic structure using both laser micromachine and soft lithography technique [6].

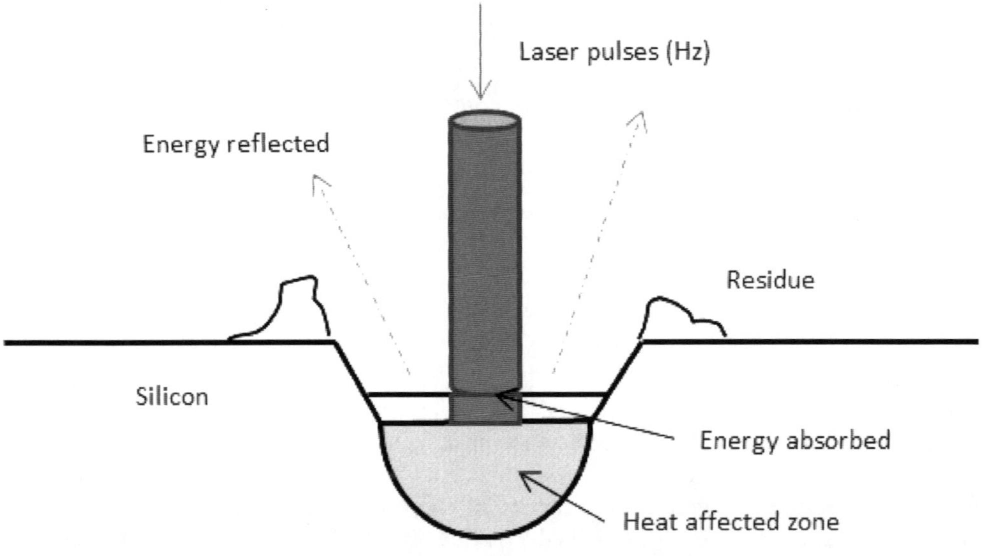

Figure 1. Laser-material interaction [7].

2. Methodology

Three types of cantilever design are initially drawn using AutoCAD software. The cantilever design are rectangular, T-shaped and triangular and the cantilever structure was designed in $4mm^2$ area. These cantilever design were created on a surface of silicon wafer with the thickness of 100 um. Initially, silicon wafer with the diameter and thickness of 100 mm and 550 μm respectively is cut into 20 mm x 20 mm dimension using a wafer scriber. Next, the wafer is dipped in the buffered oxide (BOE) solution for 30 to 60 seconds. This is to eliminate native oxide on the surface of the wafer. After that, the surface of silicon wafer is rinsed with deionized (DI) water to remove the BOE residue on the surface. The wafer is then placed on a spinner (30 seconds and 3500 rpm) for drying purposes. In order to thin the wafer to the desired thickness which is below 100 μm, the wet etching process have to be done. The thickness is chosen based on the application of the microcantilever and that it is the approximation for the optimum thickness of microcantilever [8]. Mixture of Potassium hydroxide (KOH) and isopropy alcohol (IPA) solution is prepared for wet etching purposes. The thinning process is conducted based on different parameters as depicted in Table 1, where the solution concentration and its baking temperature are varied. The etching time is fixed for one hour.

For laser micromachining process, RapidX-250 series KrF excimer is used to pattern/etch directly the microcantilever on the wafer. The machine consists of a microscope camera that is used to point the sample on the stage. This enable the sample to be monitored live during the etching process. The stage where the sample is put is a moveable three-axis in X, Y and Z direction. In order to control the laser, a specific software called Laser Control 4 Valve is used. The software includes Atlex control software (ATL), which functions are to stabilize the pulse energy or average power, repetition rate adjustment, high voltage adjustment, trigger modes, auto gas exchange, menu guided mirror exchange, transport & storage function, menu guided gas cylinder exchange, and safety interlock functions.

When using laser micromachine, there are five primary parameters that have to be taken into consideration. They are number of pulse, laser energy, number of laser shots and rectangular variable arperture (RVA) in both x- and y- direction. These parameters were varied to a certain range based on our previous work [9]. Energy (Joule) is directly directed to the volume of the excimer laser gas charging voltage. RVA-X (Rectangular Variable Aperture-X) is the x-axis of the lens in the Rapid X laser micromachining and RVA-Y (Rectangular Variable Aperture-Y) is the x-axis of the lens in the Rapid X laser micromachining. Number of shot is the number of laser shot required to obtain the desired microstructure, while number of pulse is the number of laser pulses exposed onto the wafer surface leading to ablation effect.

The silicon wafer is placed on the wafer stage and is adjusted to the center of the laser lens. Then to avoid contamination on the sample, the exhaust gas is opened to clear the dust on the surface of the silicon wafer. Next, BobCAD software is opened and the design of desired microcantilever from AutoCAD is loaded.

3. Results and discussion

During the wet etching process, the concentration of the KOH solution and its temperature were varied. The concentrations were varied at 20%, 25% and 30%, while for the temperature were increased from 60°C, 80°C and 100°C. At the end of the etching process, the result of the wafer surface should be low in surface roughness and the etch rate should be low approximately 1μm/min. The etch rate is the time taken for the silicon wafer to etch. The wafer thickness prior to wet etching is 550 μm.

Referring to the high power microscope images of wafer surface (Table 1), the higher the temperature of the solution, the roughness will be increased also. But for 25% and 30% KOH concentration, at 100°C etching process temperature, the silicon wafer was completely etched during 60 minutes. So, the etching process for them is carry out at 30 minutes only. It can be concluded that with higher KOH concentration and higher KOH solution temperature, the etch rate will be higher and it will thin the silicon wafer faster. Even though it is beneficial when the time taken for the etching process is faster but they will give a rougher wafer surface.

Table 1. Images of substrates after wet etching process at different concentrations and temperatures.

	60°C	80°C	100°C
20%		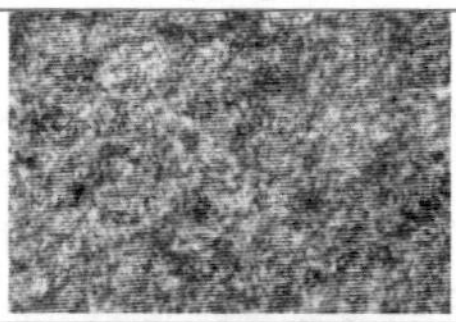	
25%	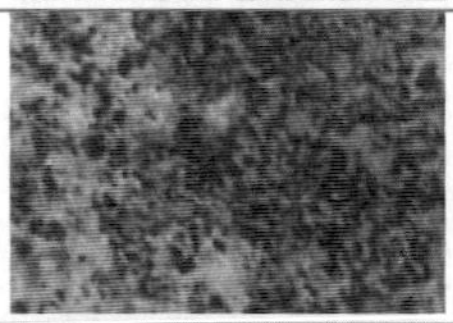		
30%			

The best parameter for etching process should be low etch rate approximately 1μm/min and low in surface roughness. From the parameter obtained, the relationship between the temperature and the etch rate can clearly be seen in the graph of temperature and etch rate as depicted in Figure 2. From the graph below, the result can be summarized that with higher temperature, etch rates will be higher also. Therefore, from the result obtained of the etch rate and the surface roughness of the wafer surface, at 80°C will give the best result for the wafer thinning at 30% KOH concentration because it will give low etch rate and less surface roughness.

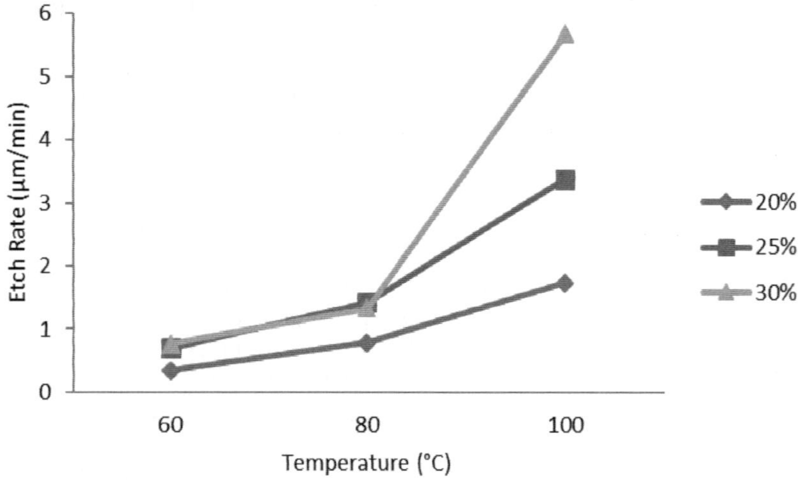

Figure 2. Etching rate of substrate with respect to temperature for different concentration

During the fabrication process using laser micromachine, the operational parameter of the machine were varied according to [8]. The number of pulse used was 300, laser energy was set to 15mJ, and the frequency, RVA-X and RVA-Y were set to 400 Hz, 5mm and 5mm respectively. The laser energy and laser pulse rate gave direct impact to the etching depth while the rectangular variable aperture (RVA) either "X" or "Y" gave direct effect to the cantilever surface dimension, or simply the width. As expected the etched area increase with laser energy, as at higher laser power, a larger part of the focused femtosecond beam has energy exceeding the machining threshold.

We found that the number of laser pulses do not have any significant effect on the etching performance of the structure. We also established that, when the pulse rate is higher, the ablated surface will be smoother.

The optimized laser micromachined parameter used in previous works to fabricate different materials are summarized in Table 2. It can be seen that the laser parameter of the micromachine depends highly on the ablated materials. The material parameters include its type and thickness, substrate reflectivity and density, and its spectral absorption. For instance, the laser energy required for the fabrication of PDMS is very low compared to the energy required to micromachine other materials. This is because the elastic properties of PDMS is lower compared to other materials. If the laser energy used is low, the frequency of the laser micromachine has to be increased. In order to effectively fabricate transparent materials with femtosecond laser pulses, the laser pulse energy must be higher than the damage threshold of the ablated material.

Table 2. Laser parameters of previous works

Frequency (Hz)	Laser energy (J)	Material
250k	2.4u	Silica [9]
250k	3u	Glass [9]
100	1.2-1.6u	PDMS [9]
100	12-40	Silicon [10]
100	12-47.5	PZT [10]
100	12-20	Pyrex [10]
400	15m	Silicon [This work]

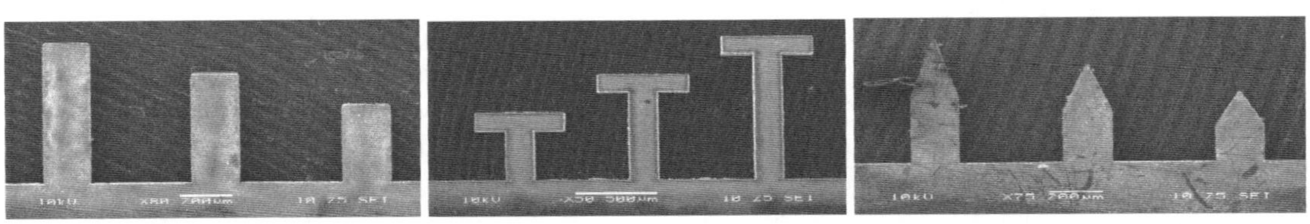

Figure 3. SEM images of rectangular, T-shaped and triangular microcantilver array

The fabricated cantilevers were inspected and measured using SEM and High power microscope, as shown in Figure 3. It can be seen from the SEM micrographs that the edge of the microcantilever structure is not completely clean as several remains and recast can be observed around the edge of the microcantilever. The remains are the materials produced throughout the ablation process, and the vaporization of the targeted material fall back onto the target substrate and accumulated around the ablation area. According to [7], these remains can be reduced by cleaning the ablated area immediately after ablation or alternatively, gas jets can also be used to remove it. This however will increase the processing time of the laser micromachining. A different suggested method to eliminate the remains is to coat the ablated area with photoresist prior to laser micromachining process. Then, photoresist developer will be used to remove the photoresist. This somehow is opposing the laser micromachine process, where the additional steps of etching should be avoided.

4. Conclusion

Laser micromachining process have the potential to contribute significantly to the development of MEMS devices. In addition, the ability of manipulating materials could also provide further improvement in the field of MEMS packaging and assembly. We have demonstrated the process of fabricating three different types of microcantilever namely rectangular, T-shaped and triangular using laser micromachine. The results were observed using high power microscope and SEM. Initially silicon substrate was etched in order to obtain the desired thickness of the cantilever. It was found that optimized etch rate is approximately 1µm/min, which yield in low surface roughness. Future works include performing experiment to measure the stiffness of the fabricated microcantilever and also to investigate the bending effect of the cantilever by applying input force and measuring its displacement.

Acknowledgments

The authors would like to thank the technicians at AMBIENCE laboratory and SoME cleanroom for their assistance. This work is supported by Research Grant UNIMAP FRGS 9003:00405.

References

[1] Li J and Ananthasuresh G K 2000 J. Micromech. Microeng. **11** 38-47
[2] White Y V, Parrish M, Li X, Davis M D and Hofmeister W 2008 *Proc. Of SPIE* vol 7039
[3] Tay F E H, van Kan J A, Watt F and Choong W O 2001 J. Micromech. Microeng. **11** 27-32
[4] Desbiens J-P, Masson P 2007 Sensors and Actuators A **136** 554-563
[5] Joohan K and Xianfan X 2003 Journal of Laser Applications **15** 4
[6] Johari S, Omar M N A and Nor M N M 2016 *Proc: Int. Conf. on Electronic Design (Phuket)*(IEEE)
[7] Mazalan M, Johari S, Ng B P and Wahab Y, *Proc. Int. Conf. on Semiconductor Electronics* (Kuala Lumpur) (IEEE)
[8] Zainol M Z, Johari S, Hamzah F, Anuar A F M, Wahab Y, Mazalan M 2015 *Proc. Regional Symposium on Micro and Nanoelectronics* (Terengganu)(IEEE)
[9] Yelena V. White, Matthew Parrish, Xiaoxuan Li, Lloyd M. Davis, William Hofmeister, 2008, Proc. of SPIE Vol. 7039 70390J-1
[10] Jean-Philippe Desbiens, Patrice Masson, 2007 Sensors and Actuators A 136, 554–563

IOP Conf. Series: Materials Science and Engineering **340** (2018) 012011 doi:10.1088/1757-899X/340/1/012011

Surface Topology and Optical Properties of Nanostructured Zinc Oxide Thin Films Prepared Using Two-Stage Solution Immersion Method

S.A. Saidi[1], M.H. Mamat[1,2,*], A.S. Ismail[1], M.F. Malek[1,2], M.M. Yusoff[1], W.R.W. Ahmad[1], N.D.Md. Sin[1], A.S. Zoolfakar[1], Z. Khusaimi[2], M. Rusop[1,2],

1)NANO-ElecTronic Centre (NET), Faculty of Electrical Engineering, Universiti Teknologi MARA (UiTM), 40450 Shah Alam, Selangor, Malaysia

2)NANO-SciTech Centre, Institute of Science, Universiti Teknologi MARA, 40450 Shah Alam, Selangor, Malaysia

*Corresponding e-mail: mhmamat@salam.uitm.edu.my

Abstract. Nanostructured zinc oxide (ZnO) films were prepared using two-stage immersion process. The properties of the prepared films were investigated using atomic force microscopy and ultraviolet-visible spectroscopy. The root-mean-square surface roughness of the nanostructured ZnO film was 79 nm. The average optical transmittance of the film was 72% in the visible spectrum. The prepared nanostructured ZnO film also has high ultraviolet absorbance.

1. Introduction

Recently, research on metal oxide nanostructured semiconductors shows fascinating results due to the unique physical and chemical properties of nanomaterials. Among the metal oxide nanostructure that has been extensively studied is zinc oxide (ZnO). The ZnO nanostructures have been applied for various electronic devices such as sensors, solar cells, and field emission devices owing to their distinctive structures and tunable conductivity and properties [6, 13, 14]. Various methods have been used to fabricate ZnO nanostructures including hydrothermal [1], chemical vapor deposition (CVD) [15], electron beam evaporation [5], radio frequency (RF) magnetron sputtering [9], thermal oxidation [2], and sol-gel methods [4].

In this paper, we reported a solution immersion method to synthesize nanostructured ZnO films on the glass substrates. The films were prepared through two stages of immersion process. The first stage of immersion process is to produce ZnO nanorod arrays, while the second stage of immersion process is to deposit nanostructured ZnO on ZnO nanorod arrays. The motivation of this research is to produce higher surface area through the introduction of ZnO nanostructures on nanorods. In addition, the study also to seek novel nanostructures configuration, which could improve electronic device performance particularly humidity sensor. The use of ZnO nanostructures such as assembled nanoflakes, nanograin, or nanoplatelets structure is likely to improve structural-mechanical stability in comparison to one-dimensional (1D) and zero-dimensional (0D) nanoparticles structures [7].

2. Experimental procedure

2.1. Preparation of First ZnO seed layer

The preparation of solution for the first ZnO seed layer used zinc acetate dihydrate ($Zn(CH_3COO)_2.2H_2O$; 99.5% purity; Aldrich), monoethanolamine (MEA, $H_2NCH_2CH_2OH$; 99.5% purity; Aldrich), aluminum nitrate nanohydrate ($Al(NO_3)_3.9H_2O$; 98% purity; Analar), and 2-methoxyethanol as a precursor, stabilizer, dopant source, and solvent, respectively. The solution was prepared at 0.4 M. The solution was stirred and heated at 80 °C for 3 hours before aged for 24 h at room temperature. The prepared homogeneous solution was spin-coated on glass substrates at 3000 rpm for 1 min. Then, the film-coated substrates were heated and dried at 150°C for 10 min. Subsequently the thin films were annealed for 1 h at 500 °C in a furnace after the procedure was repeated for a five times to increase the thickness.

2.2. Preparation of ZnO Nanorods Growth

The solution for ZnO nanorod was prepared using zinc nitrate hexahydrate ($Zn(NO_3)_2.6H_2O$; 98.5% purity; Schmidt), hexamethylenetetramine (HMT, $C_6H_{12}N_4$; 99% purity; Aldrich), and aluminum nitrate nanohydrate ($Al(NO_3)_3.9H_2O$; 98% purity; Analar),as a precursor, a stabilizer, and a dopant source, respectively. The solution was prepared at 0.1 M. The solution were prepared by mixing and dissolving the reagents into deionized (DI) water and then sonicated at 50 °C for 30 min. Then, the solutions were aged at room ambient for 2 h. Next, the solutions were poured into Schott bottles; the seed layer-coated glass substrates were placed at the bottom of the bottles. The bottles were then immersed in a water bath immersion tank at 95 °C for 2 hour. Subsequently, the obtained films were cleaned with DI water and dried at 150 °C for 10 min. Finally, the films were annealed at 500 °C for 1 h after the thin films

2.3. Preparation of Second ZnO seed layer

For the preparation of solution for the second ZnO seed layer, zinc acetate dihydrate (Zn(CH3COO)2.2H2O; 99.5% purity; Aldrich), mono-ethanolamine (MEA, $H_2NCH_2CH_2OH$; 99.5% purity; Aldrich), aluminum nitrate nonahydrate ($Al(NO_3)_3.9H_2O$; 98% purity; Analar), and 2-methoxyethanol were used as a precursor, stabilizer, dopant source, and solvent, respectively. The solution was stirred and heat at 80 °C for 3 hours before aged for 24 h at room temperature. The prepared homogeneous solution was spin-coated on ZnO nanorod array-coated glass substrate at 1000 rpm for 1 min. Next, the coated nanorod films were heated and dried at 150°C for 10 min. Finally, the samples were annealed for 1 h at 500 °C.

2.4. Preparation of ZnO nanostructures on ZnO nanorod arrays

The solution was prepared using zinc nitrate hexahydrate ($Zn(NO_3)_2.6H_2O$; 98.5% purity; Schmidt), hexamethylenetetramine (HMT, $C_6H_{12}N_4$; 99% purity; Aldrich), and aluminum nitrate nanohydrate ($Al(NO_3)_3.9H_2O$; 98% purity; Analar),as a precursor, a stabilizer, and a dopant source, respectively. The solutions were prepared by mixing and dissolving all the reagents into DI water. Then, the solution was sonicated at 50 °C for 30 min and aged at room temperature for 2 h. Next, the solutions were poured into Schott bottles; the coated nanorod films on glass substrates were placed at the bottom of the bottles. The bottles were immersed into water bath immersion tank at 95 °C for 2 h. The obtained nanostructured films were cleaned with DI water and heated at 150 °C for 10 min. Next, the obtained nanostructure films were annealed at 500 °C for 1 h.

2.5. Characterization

The surface topology of nanostructured ZnO films were characterized using atomic force microscopy (AFM) and the optical properties of the samples characterized using ultraviolet-visible (UV-Vis; Varian Cary 5000) spectrophotometer.

3. Result and Discussion

Fig. 1 shows surface topology of nanostructured ZnO film prepared using solution method. Fig.1 (a) and (b) show two-dimensional (2D) and three-dimensional (3D) AFM images of nanostructured ZnO film, respectively. The images were obtained using contact mode AFM. The scanning area of the images was 10 μm x 10 μm. The figures reveal that the sample surface is relatively rough. The measured value of root-mean-square (RMS) surface roughness was 79 nm. Our previous experiment showed that after the first immersion process, the nanorod were grown uniformly on the glass substrate [8, 10, 11]. However, after the second immersion process, the nanorod structure was not observed in the AFM images. This condition was due to the secondary growth of ZnO nanostructures on ZnO nanorod arrays after second time immersion process. These nanostructures can be observed clearly in the AFM images as shown in the circle in Fig. 1(a). In addition, the coating process of second ZnO seed layer on the nanorod arrays, which deposited on top of the nanorods, may hinder the visibility of the nanorod structure. The high surface roughness of deposited films also influenced by the chemical species presence in the films and the physical aspects during the deposition process.

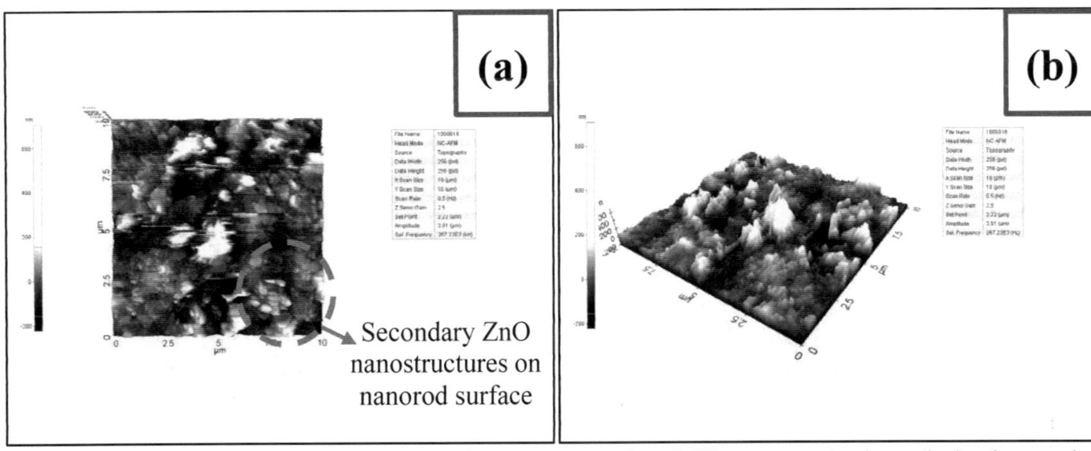

Fig. 1. (a) 2D and (b) 3D AFM images of nanostructured ZnO film prepared using solution immersion method.

Fig 2 shows the optical measurement results of nanostructured ZnO film. The results for ZnO nanorod arrays coated with second ZnO seed layer was also shown for comparison. Fig. 2(a) displays the transmittance spectra of nanostructured ZnO film at wavelength of 350 to 800 nm. The films showed good transmittance properties in the visible region (400-800 nm) with average transmittance approximately 72 %. This transmittance value was slightly lower than that of the ZnO nanorod arrays coated with second ZnO seed layer (78 %). This condition was due to decrease in void of the film after second time immersion process [3]. Furthermore, the scattering of light might be prevalent in the nanostructured ZnO film due to rough surface induced by secondary nanostructured ZnO growth [12].

Figure 2(b) shows the absorbance spectra of nanostructured ZnO film and film of ZnO nanorod arrays coated with second ZnO seed layer. The results show that both films have high absorbance properties in the UV region below 400 nm. However, the nanostructured ZnO film possess higher light absorbability in the UV region. This condition might be due to light scattering effects as discussed previously. The light scattering phenomenon results in longer transmission path of the light in the film. This condition enables the light absorbance at the optimum capacity by nanostructured ZnO due to light traveling across more ZnO surfaces. In addition, the higher fraction of nanostructured ZnO film as compare to film of ZnO nanorod arrays coated with second ZnO seed layer enables UV light to be absorb efficiently. It also can be observed that the ZnO films shows a cut-off wavelength at 380 nm. The band gap energy of ZnO were reported to be in the region between 3.2 to 3.3 eV, which matches with cut-off wavelength of nanostructured ZnO in this study. The cut-off wavelength observed through UV-vis spectroscopy result can be as a preliminary tool for material bandgap estimation.

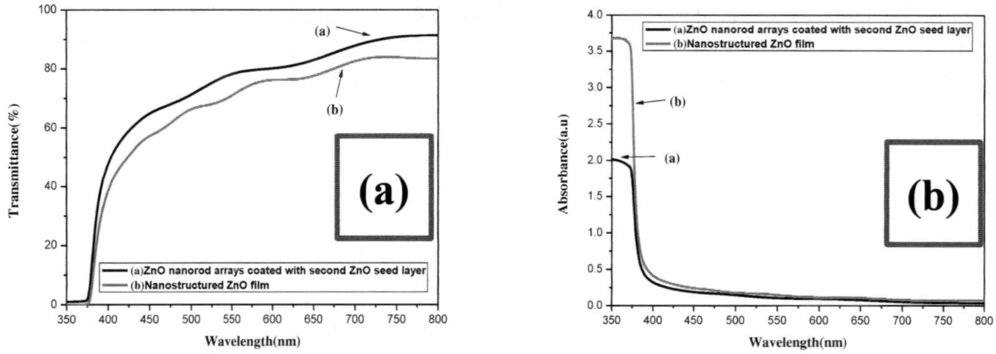

Fig. 2. (a) Optical transmittance and (b) absorbance spectra of nanostructured ZnO film and film of ZnO nanorod arrays coated with second ZnO seed layer.

4. Conclusion

In summary, nanostructured ZnO film have been successfully synthesized by two-stage immersion method on glass substrate. The structural and optical properties of the film were investigated using AFM and UV-vis spectroscopy, respectively. The nanostructured ZnO film has high surface roughness with RMS roughness value of 79 nm. The optical transmittance spectra revealed that nanostructured ZnO films exhibit high transparency in the visible region with average transmittance of 72 %. Moreover, the absorbance of nanostructured ZnO film in the UV region below 400 nm improved compare to that of film of ZnO nanorod arrays coated with second ZnO seed layer. These results due to light scattering effect induced by rough surface area and the existence of secondary ZnO nanostructures on the nanorod surfaces.

5. Acknowledgement

The authors also would like to thank the Faculty of Electrical Engineering, Universiti Teknologi MARA and the Ministry of Higher Education of Malaysia for their financial support of this research. This work was supported by the Fundamental Research Grant Scheme (600-RMI/FRGS 5/3 (57/2015)) from the Ministry of Higher Education of Malaysia.

6. References

[1] Dou Y, Wu F, Mao C, Fang L, Guo S and Zhou M 2015 Enhanced photovoltaic performance of ZnO nanorod-based dye-sensitized solar cells by using Ga doped ZnO seed layer *Journal of Alloys and Compounds* **633** 408-14

[2] Florica C, Preda N, Costas A, Zgura I and Enculescu I 2016 ZnO nanowires grown directly on zinc foils by thermal oxidation in air: Wetting and water adhesion properties *Materials Letters* **170** 156-9

[3] Geunjae K, Sungmook J and Kijung Y 2011 Multifunctional transparent ZnO nanorod films *Nanotechnology* **22** 115705

[4] Hendi A A, Alorainy R H and Yakuphanoglu F 2014 Humidity sensing characteristics of Sn doped Zinc oxide based quartz crystal microbalance sensors *Journal of Sol-Gel Science and Technology* **72** 559-64
[5] Kim I S, Jeong E-K, Kim D Y, Kumar M and Choi S-Y 2009 Investigation of p-type behavior in Ag-doped ZnO thin films by E-beam evaporation *Applied Surface Science* **255** 4011-4
[6] Kuo S-Y and Lin H-I 2014 Field emission characteristics of zinc oxide nanowires synthesized by vapor-solid process *Nanoscale Research Letters* **9** 70
[7] Liang Y, Guo N, Li L, Li R, Ji G and Gan S 2015 Fabrication of porous 3D flower-like Ag/ZnO heterostructure composites with enhanced photocatalytic performance *Applied Surface Science* **332** 32-9
[8] Mamat M H, Che Khalin M I, Nik Mohammad N N H, Khusaimi Z, Md Sin N D, Shariffudin S S, Mohamed Zahidi M and Mahmood M R 2012 Effects of Annealing Environments on the Solution-Grown, Aligned Aluminium-Doped Zinc Oxide Nanorod-Array-Based Ultraviolet Photoconductive Sensor *Journal of Nanomaterials* **2012** 189279
[9] Mamat M H, Hafizah N N and Rusop M 2013 Fabrication of thin, dense and small-diameter zinc oxide nanorod array-based ultraviolet photoconductive sensors with high sensitivity by catalyst-free radio frequency magnetron sputtering *Materials Letters* **93** 215-8
[10] Mamat M H, Ishak N I, Khusaimi Z, Zahidi M M, Abdullah M H, Muhamad S, Sin N D M and Mahmood M R 2012 Thickness-Dependent Characteristics of Aluminium-Doped Zinc Oxide Nanorod-Array-Based, Ultraviolet Photoconductive Sensors *Japanese Journal of Applied Physics* **51** 06FF3
[11] Mamat M H, Khusaimi Z, Zahidi M M, Suriani A B, Siran Y M, Rejab S A M, Asis A J, Tahiruddin S, Abdullah S and Mahmood M R 2011 Controllable Growth of Vertically Aligned Aluminum-Doped Zinc Oxide Nanorod Arrays by Sonicated Sol–Gel Immersion Method depending on Precursor Solution Volumes *Japanese Journal of Applied Physics* **50** 06GH4
[12] Mamat M H, Khusaimi Z, Zahidi M M, Suriani A B, Yusop M Z, Tanemura M and Mahmood M R 2012 Effects of Aluminium Doping and Electrode Distance on the Performance of Aligned Zinc Oxide Nanorod Array-Based Ultraviolet Photoconductive Sensors *Japanese Journal of Applied Physics* **51** 06FE4
[13] Narimani K, Nayeri F D, Kolahdouz M and Ebrahimi P 2016 Fabrication, modeling and simulation of high sensitivity capacitive humidity sensors based on ZnO nanorods *Sensors and Actuators B: Chemical* **224** 338-43
[14] Rouhi J, Mamat M H, Ooi C H R, Mahmud S and Mahmood M R 2015 High-Performance Dye-Sensitized Solar Cells Based on Morphology-Controllable Synthesis of ZnO–ZnS Heterostructure Nanocone Photoanodes *PLoS ONE* **10** e0123433
[15] Sáenz-Trevizo A, Amézaga-Madrid P, Pizá-Ruiz P, Antúnez-Flores W, Ornelas-Gutiérrez C and Miki-Yoshida M 2015 Single and multi-layered core-shell structures based on ZnO nanorods obtained by aerosol assisted chemical vapor deposition *Materials Characterization* **105** 64-70

IOP Conf. Series: Materials Science and Engineering **340** (2018) 012012 doi:10.1088/1757-899X/340/1/012012

Direct growth of ZnO tetrapod on glass substrate by Chemical Vapor Deposition Technique

M. F. M. Fadzil, R. A. Rahman, N. E. A. Azhar, T. N. T. A. Aziz, Z. Zulkifli*

Nano-ElecTronic Center, Faculty of Electrical Engineering, UiTM Shah Alam

*zurita101@salam.uitm.edu.my

Abstract. This research demonstrates the growth of ZnO tetrapod structure on glass substrate for different types of flow gas and at different growth temperatures. The study on the morphological structure and electrical properties of ZnO thin film growth by Chemical Vapour Deposition (CVD) technique showed that the optimum growth temperature was obtained at 750°C with ZnO nanotetrapod morphological structure. Introducing Nitrogen gas flow during the growth process exhibited leg-to-leg linking ZnO tetrapods morphology. The electrical properties of ZnO tetrapods film were measured by using two point probes and it shows that, the sample growth in Ar and O_2 atmosphere have better I-V characteristic.

1. Introduction

Zinc oxide (ZnO) is known has a good characteristic on electrical properties and can used for higher temperature condition, good transparency and high electron mobility [1-3]. In addition, ZnO are one of the several metal oxide that can be used as a fabrication for the others generation of optoelectronic device and display device. ZnO has a larger bandgap of 3.37 eV. The ultraviolet light emitting diode and lasers are attracted with characteristic of zinc because it has a good wide band gap [4]. ZnO can growth in various types of morphology structure such as nanowires, nanorods, nanobelt, nanosring, nanostars, nanotube, nanopropellers and others. ZnO nanostructure can be growth from several deposition techniques such as vacuum, thermal evaporation, dip-coating, sol gel and other [5]. The chemical vapour deposition (CVD) has been chosen in this experiment due to its simple preparation and low cost deposition method. Different gas flow, growth time and annealing time may affect the properties of ZnO either in morphology or electrical characteristic. Thus, in this experiment, it is essential to study the effect of different gas flow on the growth of ZnO tetrapod as promising films for any electronic applications.

The structural properties of ZnO have been observed using field-emission scanning electron microscopy (FESEM) and the electrical properties was characterized using I-V measurement.

2. Experimental details

The ZnO tetrapods were prepared by thermal evaporation of 2g Zn powder with a purity of 99.9%. Glass slides were used as the substrate and were cut in the size of 2.5cm x 2.5cm. Prior to the experiment, glass substrates were cleaned using acetone, methanol and deionized water. Glass substrates were immersed in a beaker with acetone and followed by methanol and deionized water, respectively. The ultrasonic machine was set to 45°C and takes 15 minutes for each solution. The cleaning process is conducted because to avoid from nanoparticles or contamination of small structure form on the surface of the glass substrate. Small particle that adhere on the glass surface are feared can scratch the surface during the cleaning process. If any scratches formed on the glass surface, it can effect on surface morphology of the sample.

The Chemical Vapor Deposition (CVD) has two furnaces. The growth temperature at the first furnace was heated at 650 °C, 750 °C and 850 °C, individually. Second furnace where the substrate was placed was kept at 500 °C for all growth temperature. Before the growth process, the tube was cleaned by allows Argon gas to flow in the rate of 68 sccm for 15 mins to eliminate unwanted particles. Once the furnace reach the desired temperature, Argon gas was maintained flow for 30 mins then followed by Oxygen gas flow at 5 sccm for another 30 mins. At this point, Argon and Oxygen was mixed in the tube. The evaporated Zn vapor was reacted with oxygen in air to form the tetrapod-shaped ZnO crystals. After 30 mins, the furnace was turned off and cooled to room temperature. Then, the samples were collected for characterization. This method was repeated for samples with Nitrogen gas flow at 13 sccm.

The morphology of as-grown and annealed samples was observed by field-emission scanning electron microscopy (FE-SEM). Energy dispersive X-Ray (EDX) was used to analyze the element in the fabricated film. The electrical properties of the samples were characterized by two-point probe solar simulator. Gold was sputtered on the sample as the metal contact.

3. Results and Discussion

The density of as-grown and annealed ZnO tetrapods structure and the electrical properties were investigated for all growth temperature. The flow of nitrogen gas during the growth process shows effect on the density of the ZnO tetrapods.

3.1. Surface Morphology

The FESEM images of direct-grown ZnO tetrapods growth at 650 °C, 750 °C and 850 °C on glass substrate are shown in Figure 1 (a)-(c). All growth temperatures show formation of ZnO tetrapod on the glass substrate. The density of ZnO tetrapods covers most of the surface area of the glass substrate. The density at 650 °C is less than ZnO tetrapod during growth at 850 °C. The evaporation of ZnO powder is increasing when higher growth temperature is used. Inset figure in Figure 1 (a) shows the ZnO tetrapods with an average length of 1.35 μm and 0.08 μm width. When the growth temperature was set to 750 °C, ZnO tetrapods width and length shows an increases in average of 0.15 μm and 1.69 μm, respectively as shown in inset figure of Figure 1 (b). While, inset figure of Figure 1 (c) shows the higher the growth temperature, the longer length of ZnO tetrapods but no changes on the width of ZnO tetrapods. However, higher growth temperature tends to bend the branch of ZnO tetrapod which could increase the resistance due to the bending structures. ZnO tetrapods growth at 650 °C shows a large boundaries between each other compared to 750 °C. In the next steps, we introduce N_2 flow during the growth process to observe the effect on ZnO tetrapod at the same growth temperatures.

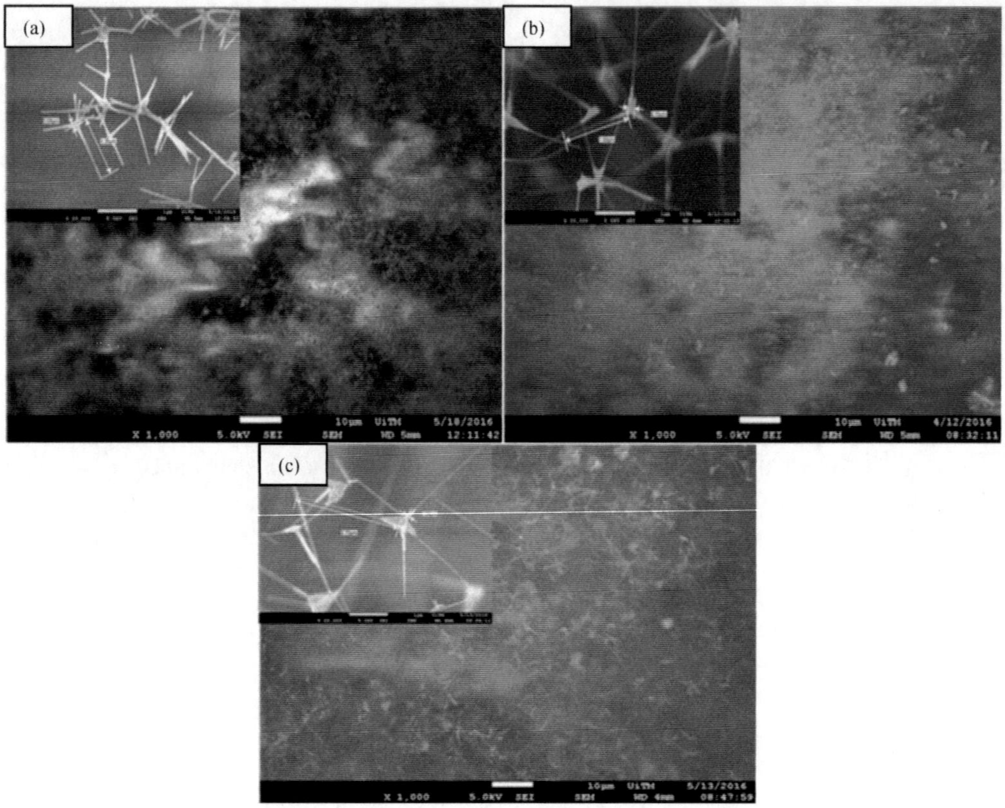

Figure 1: ZnO tetrapod growth at (a) 650 °C (b) 750 °C and (c) 850 °C

3.2. Effect of Nitrogen on ZnO tetrapods

Besides Oxygen and Argon flow, we introduced 13 sccm of N_2 flow rate during ZnO tetrapod growth. From the FESEM images as shown in Figure 2 (a)-(c) for 650 °C , 750 °C and 850 °C, respectively, Nitrogen-rich atmosphere shows abundant of ZnO tetrapods compared to only Argon and Oxygen gas. Inset figures shows higher magnification of the tetrapods. Morphology analyses showed that, nitrogen used in this process leads to an enhancement of ZnO tetrapod densification [6]. Optimum density is required to create a path for electron movements. Thus, higher density of ZnO tetrapod growth in nitrogen ambient increases the surface area which is useful to apply on any application that needs higher sensitivity such as sensors.

On the upper region, we observed leg-to-leg linking of ZnO tetrapods as shown in Figure 3. The inset figure shows higher magnification clearly shows the linking between one leg with another tetrapods leg. The morphology of linked legs is due to the existing of N_2 in the quartz tube during the growth process which increases the pressure and resulting ultrahigh supersaturation condition inside the tube. The high supersaturation condition results to an increase in the nucleation of ZnO tetrapods [7]. Formation of leg-to-leg ZnO tetrapods on the upper region can be explained due to high vapor pressure in the tube which induced more vapor accumulated on top of the tube and leading to condensation of leg-to-leg linking ZnO tetrapods.

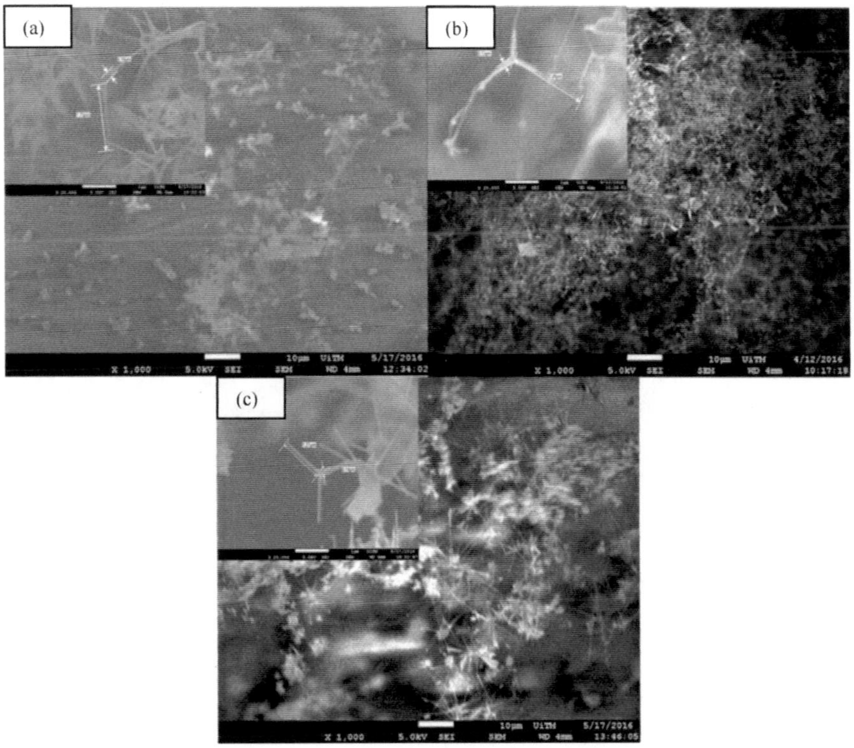

Figure 2: ZnO tetrapod growth under mixing of N_2, Ar and O_2 flow at (a) 650 °C (b) 750 °C and (c) 850 °C

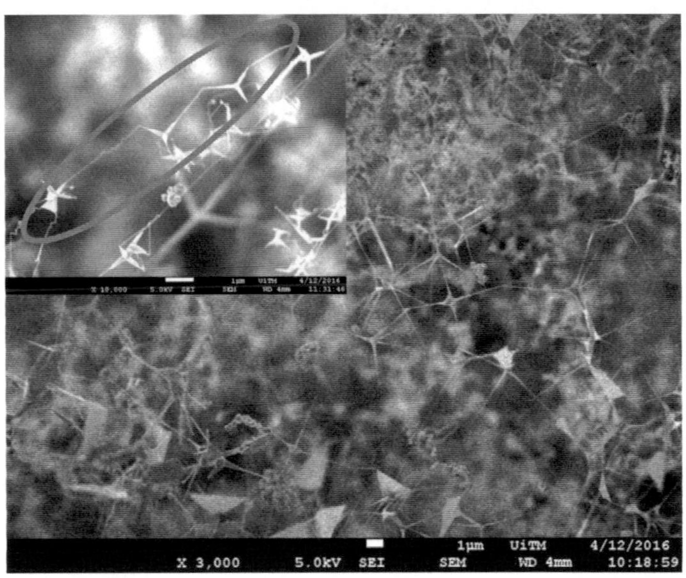

Figure 3: Leg-to-leg ZnO tetrapods on the upper region of ZnO film. Inset figure: High magnification of ZnO tetrapods leg-to-leg linking

Figure 4 shows EDX pattern of direct growth ZnO tetrapods on glass substrate. The EDX data reveals that all the samples are composed of Zn and O elements. The Si element observed from the EDX data is believed from the glass substrate.

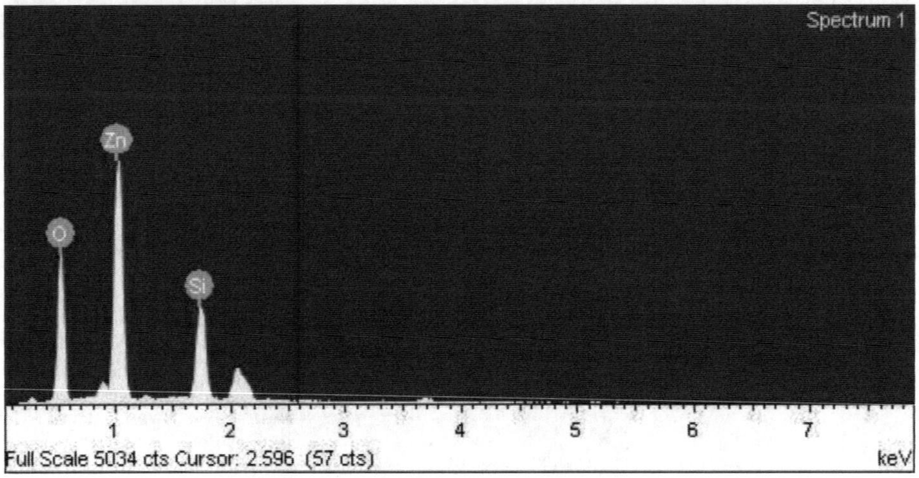

Figure 4: EDX pattern of ZnO tetrapods

3.3. Electrical properties

Figure 5 shows the I-V characteristic of ZnO tetrapods samples prepared under N_2 atmosphere and without N_2 atmosphere. From the I-V curve it shows that, sample prepared with N_2 flow resulting in lower conductivity. The reasons can be assume due to higher density by stacking morphology and leg-to-leg linking might increase the resistance throughout the path.

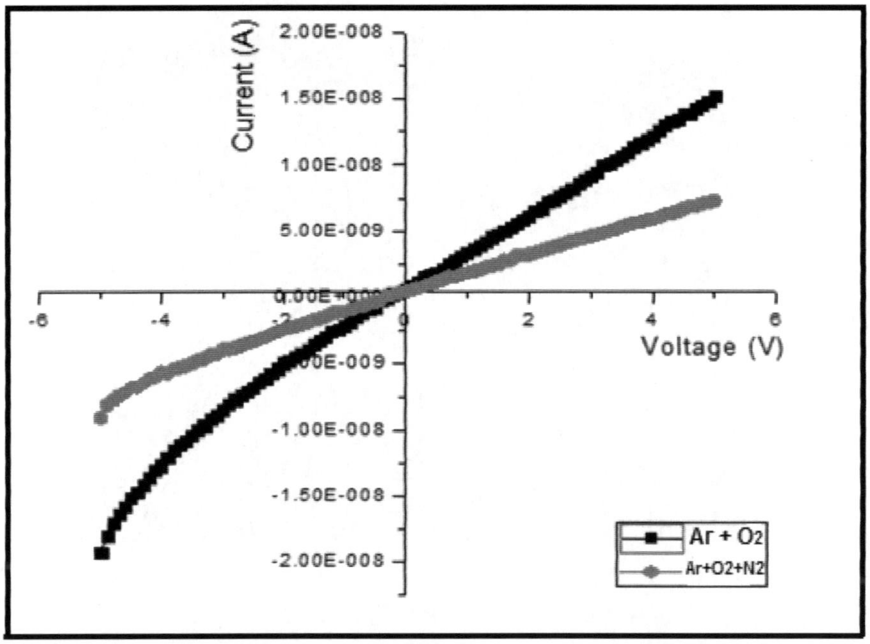

Figure 5: Electrical characteristic of ZnO tetrapods

4. Conclusions

Based from all growth temperatures, 750 °C shows the optimum growth temperature for both atmosphere conditions, with and without N_2. Higher temperature tends to bend the leg of the ZnO tetrapods thus potential to increase the resistance. Based from the morphology, lower growth temperature shows isolated ZnO tetrapods which lead to the higher needs of external voltage in order to excite the movement of electrons. The leg-to-leg linking ZnO tetrapods is beneficial for electronic, photonic, opto-electronic, and sensing applications due to higher surface area.

Acknowledgment

The author would like to NET technicians and community for their guidance and support in this research. The authors would also like to thank the Ministry of Higher Education (MOHE) for the financial support through FRGS (grant number: 600-RMI/FRGS 5/3 (147/2015)).

References

[1] Mahmoud, W. E., & T. Al-Harbi, **2011**, "Synthesis, characterization and photoluminescence properties of Sn doped ZnO nanonails", *Journal of Crystal Growth*, **327**(1), 52–56.

[2] Ching-Ming Hsu , Hon-Bin Lin, Wen-Tuan Wu, **2009**, "Surface textured molybdenum zinc oxide for light diffusion enhancement", *Thin Solid Films* **517**, 3717–3720.

[3] A. Boukhachem a,n, B.Ouni a, M.Karyaoui b, A.Madani c, R.Chtourou b, M.Amlouk a, **2012** "Structural, Opto-thermal and Electrical Properties of ZnO:Mosprayed thin films", *Materials Science in Semiconductor Processing* **15** 282–292.

[4] S. Dhara and P. K. Giri, **2012**, "Stable p-type conductivity and enhanced photoconductivity from nitrogen-doped annealed ZnO thin film," *Thin Solid Films*, vol. **520**, no. 15, pp. 5000–5006,.

[5] S. Kitova, I. Kalaglarski, R. Stoimenov, R. Kazakov, and V. Mankov, **2013**, "Zinc oxide nanostructures prepared on glass substrates by thermal evaporation of zinc at atmospheric pressure," Bulg. *Chem. Commun.*, vol. **45**, no. SPEC. ISSUE B, pp. 99–104,.

[6] A Toumiat , S Achour, A Harabi, N Tabet, M Boumaour and M Maallemi, **2006**, "Effect of nitrogen reactive gas on ZnO nanostructure development prepared by thermal oxidation of sputtered metallic zinc", *Nanotechnology* **17**, 658–663.

[7] Meechai Thepnurat, Torranin Chairuangsri, Niyom Hongsith, Pipat Ruankham, and Supab Choopun, **2015** "Realization of Interlinked ZnO Tetrapod Networks for UV Sensor and Room-Temperature Gas Sensor", *ACS Appl. Mater. Interfaces*, **7** (43), pp 24177–24184

The Effect of SiC Polytypes on the Heat Distribution Efficiency of a Phase Change Memory.

M.S. Aziz, Z. Mohammed, R.I. Alip

Faculty of Electrical Engineering,
Universiti Teknologi MARA 40450 Shah Alam, Selangor

shariman_1285@yahoo.com

Abstract. The amorphous to crystalline transition of germanium-antimony-tellurium (GST) using three types of silicon carbide's structure as a heating element was investigated. Simulation was done using COMSOL Multiphysic 5.0 software with separate heater structure. Silicon carbide (SiC) has three types of structure; 3C-SiC, 4H-SiC and 6H-SiC. These structures have a different thermal conductivity. The temperature of GST and phase transition of GST can be obtained from the simulation. The temperature of GST when using 3C-SiC, 4H-SiC and 6H-SiC are 467K, 466K and 460K, respectively. The phase transition of GST from amorphous to crystalline state for three type of SiC's structure can be determined in this simulation. Based on the result, the thermal conductivity of SiC can affecting the temperature of GST and changed of phase change memory (PCM).

1. Introduction

Phase change Memory (PCM) is a non-volatile memory but there are many advantages of the phase-change memory such as high endurance, low programmable energy, fast switching speed, good data retention and excellent scalability[1] [2] [3]. The history of phase change memory has began when these device were first proposed by S.Ovshinsky in late 1960s [1]. He had reported that the chalcogenide material can be reversible change in resistivity and reflectivity when changing between amorphous state and crystalline state. The changing of PCM from amorphous state to crystalline state or vice versa allows data can be stored or deleted.

Phase change materials usually exist in an amorphous phase but one or sometime several crystalline phases. The phase change material in amorphous state has characteristic like high resistivity and short range atomic order. While, the phase change material in crystalline state has a low resistivity and long range atomic order. The changing of atom in phase change material will affect the process of amorphous and crystallization. The phase change material that used in this research is $Ge_2Sb_2Te_5$ (Germanium-antimony-tellurium, GST) because it has high thermal stability with additional of Ge, very high crystallization rate and very good reversibility between amorphous and crystalline phases. It often used in rewritable optical discs and phase-change memory applications. This material has 450K crystallization temperature and 900K melting temperature [5].

Silicon carbide (SiC) is a semiconductor material and may be heated to 600 C in operation and may have a resistivity that does not significantly go down with increasing temperature. In this research, three types of silicon carbide's structure will be used; 3C-SiC, 4H-SiC and 6H-SiC. These three structures have a different of thermal conductivity. GL Harris 1995 said the 3 types of SiC; 3C-SiC,

4H-SiC, and 6H-SiC have a different thermal conductivity which is 320 W/mK, 370 W/mK, and 490 W/mK respectively but the heat capacity is 690 J/kgK, same for all types of SiC [9]. Golberg et al.2001 said the thermal conductivity for 3C-SiC is 360 W/mK [10]. For this research, the different of thermal conductivity of silicon carbide can affect the temperature of phase change material.

In this paper, the COMSOL 5.0 Multiphysic software was used for simulating the different of structure of silicon carbide as a heater layer. In COMSOL software, the joule heating is a one of multiphysic interface is used to model resistive heating. This multiphysics interface adds a heat transfer in solid interface and electric current interface. These two elements important to produce the result that is almost equal to the actual situation.

2. Experimental

Silicon carbide (SiC) is an interesting material that has found application in a variety of industries. SiC has an exclusive properties to make SiC a perfect candidates for high power, high temperature electronic devices such as high melting point, chemical and thermal stability, high hardness and strength, and high erosion resistance[7] [8]. For this research, SIC was selected as a heater layer for ensure the GST can get enough heat for phase transition from amorphous to crystalline state. The three types of SiC;s structure that used in this simulation is 3C-SiC, 4H-SiC and 6H-SiC. Figure 1 show the different structure of SiC. These three structures have a different thermal conductivity but same heat capacity. Table 1 shows the thermal conductivity and heat capacity for three types of SiC's structure.

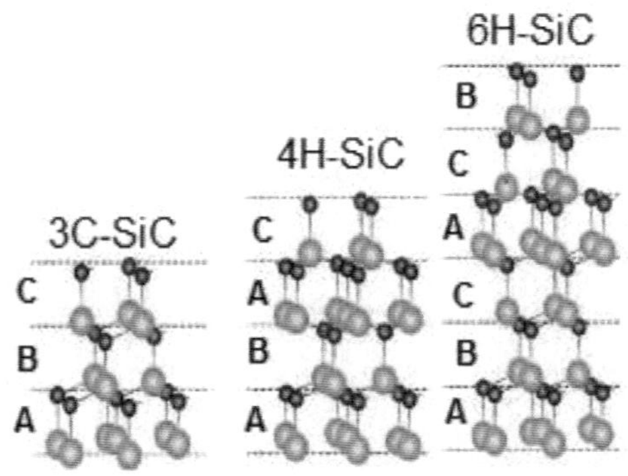

Figure 1 : The three types of SiC with different structure.

Table 1 : Thermal Conductivity and Heat Capacity of Silicon Carbide's Structure

Structure SiC	3C-SiC	4H-SiC	6H-SiC
Thermal Conductivity W/m.K	320	370	490
Heat Capacity J/kg.K	690	690	690

The separated heater structure was used in this simulation using the COMSOL 5.0 Multiphysics software. Separated heater structure consist of four electrodes, GST as memory layer, SiC as heater layer and ZnS-SiO2 as a capping layer and insulator layer. The function of capping layer is to avoid the thermal energy loss when SET process occurs. The insulator layer used to ensure the heater layer not connected electrically to the memory layer, GST. The figure 2 shows the separated heater structure of PCM [5] [6]. The voltage pulse is applied at the two top electrodes. The heater SiC will be heat the insulator. Heat will be transferred to the GST and change to the crystalline state. With this structure, the power can be controlled depend on the voltage pulse because the Rh is constant (SiC).

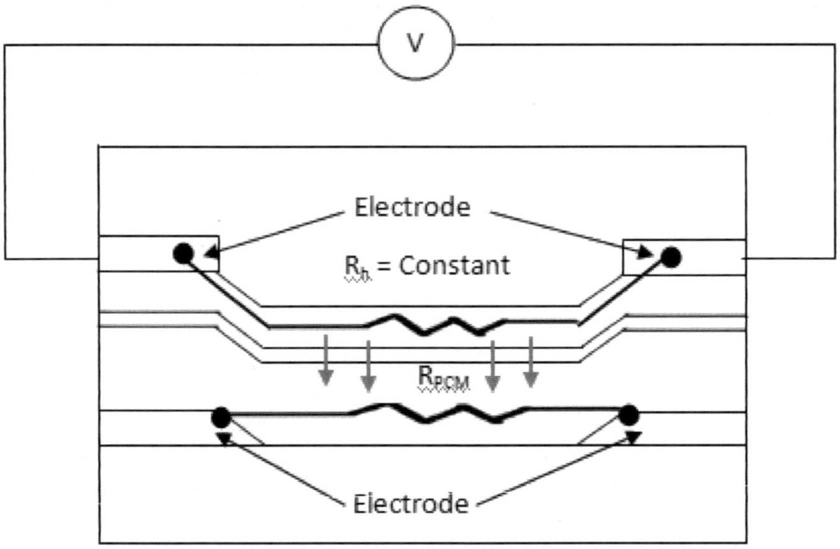

Figure 2 : The Separated heater structure of PCM.

From this simulation, the temperature of GST will be measured from 0.7V with 100ns pulse width until the GST was changed from amorphous to crystalline state. The GST will change from amorphous to crystalline state when the temperature of GST reached over 450K. This change is known as a phase transition of GST. The temperature and phase transition of GST from amorphous to crystalline state can be obtained from this simulation.

3. Result and Discussion

From the simulation, the temperature can be obtained using the three types of SiC's structure. The result of simulation can be seen in the figure 3 below when the voltage is 1.0V with 100ns pulse width. Figure 3 (a) show the temperature is 467K at GST using the 3C-SiC structure, figure 3 (b) shows the temperature is 466K at GST using the 4H-SiC structure and figure 3 (c) show the temperature is 460K at GST using the 6H-SiC structure.

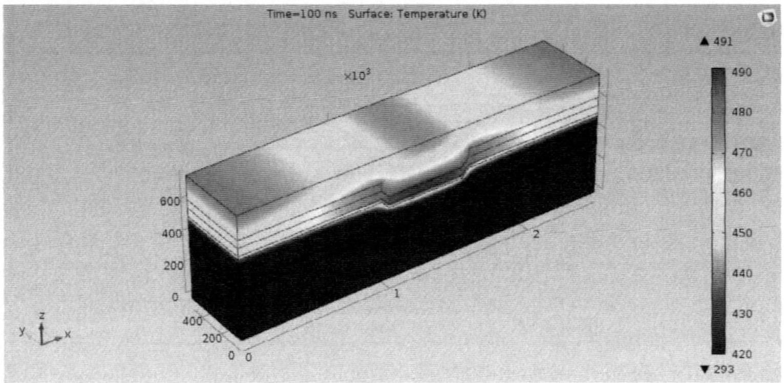

Figure 3 (a): The temperature when voltage 1.0V with 100ns pulse width was applied at 3C-SiC structure.

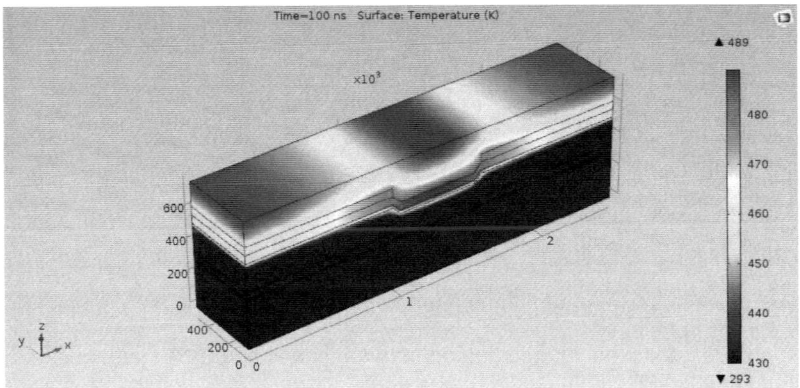

Figure 3 (b): The temperature when voltage 1.0V with 100ns pulse width was applied at 4H-SiC structure.

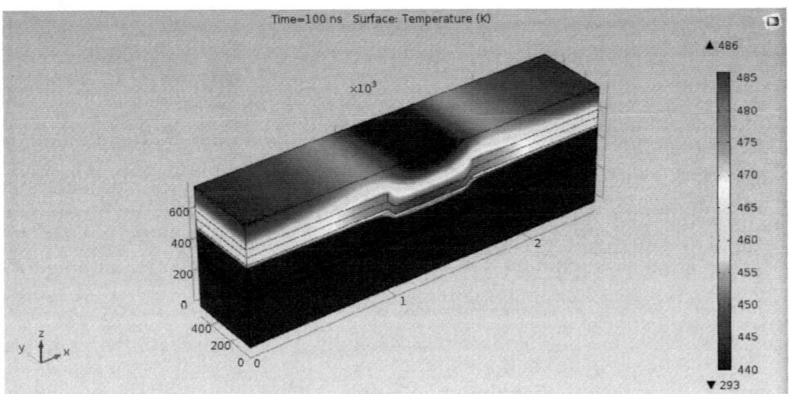

Figure 3 (c): The temperature when voltage 1.0V with 100ns pulse width was applied at 6H-SiC structure.

From the figure 3 (a), (b) and (c), the temperature of GST was decreased when the thermal conductivity of SiC was increased. The table 2 shows the temperature of GST when voltage 0.7V with 100ns pulse width was applied at the top electrode until the GST was changed to crystalline state. From this result, the thermal conductivity of SiC can affect the temperature of GST. The red color show the GST already changed to crystalline state, while the black color show the GST still in amorphous state.

Table 2 : The Temperature of GST for three type of SiC's structure.

	Temperature (K)		
Voltage (V)	**3C-SiC**	**4H-SiC**	**6H-SiC**
0.7	392.67	391.22	389.56
0.8	420.42	418.92	416.56
0.9	447.55	447.2	444.44
1.0	467.19	466.4	460.81
1.1	487.67	487.36	481.36
1.2	515.79	512.29	508.21

Figure 4 shows the phase transition of GST when voltage 0.9V and 1.0V with 100ns pulse width for three types of SiC's structure. When the voltage applied is 0.9V, the phase transition of GST will be changed from amorphous but the GST not completely changed to crystalline state. The percentage phase transition of GST for 3C-SiC, 4H-SiC and 6H-SiC are 47%, 44% and 29%, respectively. When the voltage 1.0V, the phase transition of GST for 3C-SiC changed 100% to crystalline state but for 4H-SiC and 6H-SiC the percentage phase transition are 98% and 96%, respectively. Table 3 shows summary of phase transition of GST with voltage 0.9V and 1.0V with 100ns pulse width for three types of SiC's structure.

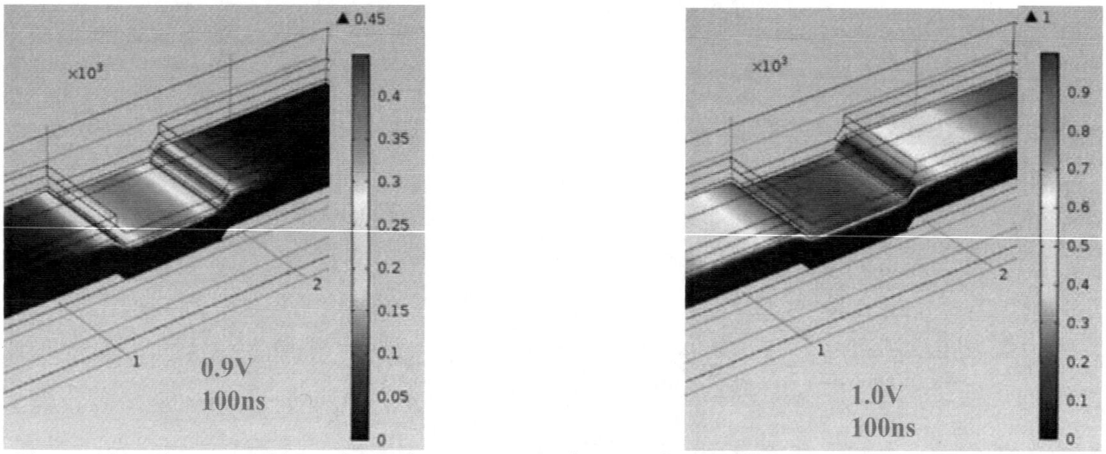

Figure 4(a): Phase transition of GST between amorphous and crystalline state when 0.9V with 100ns pulse width for 3C-SiC structure.

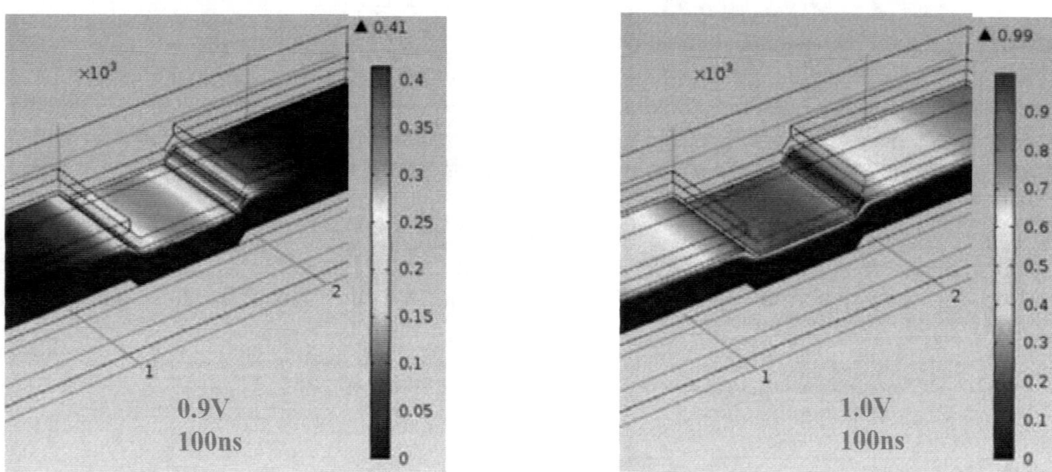

Figure 4(b): Phase transition of GST between amorphous and crystalline state when 0.9V with 100ns pulse width for 4H-SiC structure.

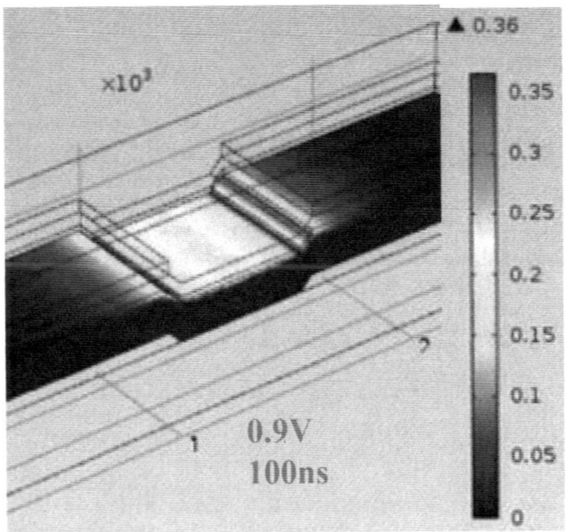

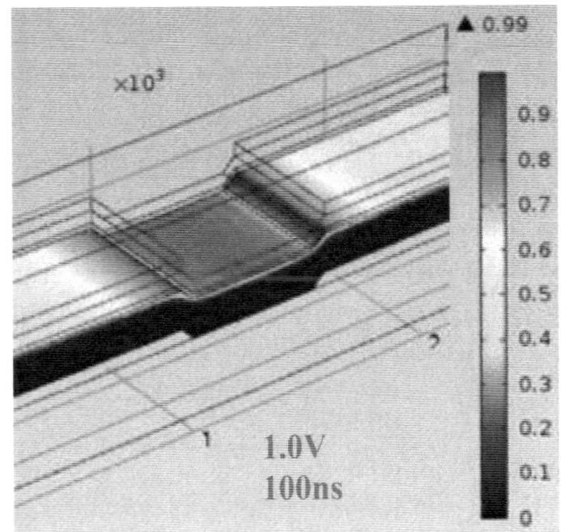

Figure 4(c): Phase transition of GST between amorphous and crystalline state when 0.9V with 100ns pulse width for 6H-SiC structure.

Table 3 : summary of phase transition of GST with voltage 0.9V and 1.0V with 100ns pulse width for three types of SiC's structure.

	Percentage of phase transition GST		
Voltage (V)	**3C-SiC**	**4H-SiC**	**6H-SiC**
0.9	47%	44%	29%
1.0	100%	98%	96%

4. Conclusion

In conclusion, the thermal conductivity of silicon carbide can affect the temperature of GST and also the phase transition of GST. The structure 3C-SiC has a lower thermal conductivity compared with the structure 4H-SiC and 6h-SiC. The lower thermal conductivity of SiC will caused the temperature of GST increased. The temperature of GST when using 3C-SiC as a heating element is 467K, while the temperature of GST when using 4H-SiC and 6H-SiC as a heating element are 466K and 460K, respectively. The phase transition of GST for 3C-SiC at 1.0V with 100ns pulse width is 100% completely changed from amorphous to crystalline state. For 4H-SiC and 6H-SiC, the phase transition of GST are 98% and 96% only but the data still can be stored at this time. Based on result simulation, the lower of thermal conductivity of SiC can increased the temperature of GST more efficient and phase transition of GST can change from amorphous to crystalline state easily.

References

[1] Hu Y, Feng X, Zhai J, Wen T, Lai T, Song S, et al. 2014 *Superlattice-like Ge 8 Sb 92 / Ge thin films for high speed and low power consumption phase change memory application.* **93** 4–7

[2] Lacaita a. L. 2006 *Phase change memories: State-of-the-art, challenges and perspectives. Solid State Electron.* **50(1)** 24–31.

[3] Wong H-SP, Raoux S, Kim S, Liang J, Reifenberg JP, Rajendran B, et al. 2010 *Phase ChangeMemory.* Proc IEEE **98(12)** 2201–27.

[4] Crespi L, Ghetti A, Boniardi M, Lacaita AL. 2014 *Electrical conductivity discontinuity at melt in phase change memory.* **35(7)** 747–9.

[5] Irma R, Kobayashi R, Zhang YL. 2013 *A Novel Phase-Change Memory with a Separate Heater Characterized by Constant Resistance for Multilevel Storage.* **534** 136–40.

[6] Yin Y, Alip RI, Zhang YL, Hosaka S. 2012 *Material Engineering for Low Power Consumption*

and Multi-Level Storage in Lateral Phase-Change Memory. **490-495** 3286–90.
[7] Daouahi M, Omri M, Ghani A, Kerm Y, Al-agel FA, Rekik N. *2015 Spectrochimica Acta Part A : Molecular and Biomolecular Spectroscopy An investigation on the effect of high partial pressure of hydrogen on the nanocrystalline structure of silicon carbide thin films prepared by radio-frequency magnetron sputtering.* **136** 1409–17.
[8] Amirthan G, Udaya A, Balasubramanian M. 2011 *Thermal conductivity studies on Si / SiC ceramic composites.* **09** 003.
[9] Harris, G.L. 1995 *Thermal conductivity of SiC*, **5-6.**
[10] Goldberg Yu., Levinshtein M.E., Rumyantsev S.L.2001 *Properties of Advanced SemiconductorMaterials GaN, AlN, SiC, BN, SiC, SiGe*, **93-148**.

Acknowledgments

This work is supported and sponsored by the Institute of Research Management and Innovation (IRMI) Universiti Teknologi Mara (Uitm) Shah Alam under Geran Insentif Penyeliaan (GIP) (Grant No. 600-IRMI/MyRA 5/3/GIP (072/2017)).

IOP Conf. Series: Materials Science and Engineering **340** (2018) 012014 doi:10.1088/1757-899X/340/1/012014

Real time fish pond monitoring and automation using Arduino

Z. Harun[1,*], E. Reda[1,2] and H. Hashim[3]

[1]Faculty of Engineering and Built Environment, Universiti Kebangsaan Malaysia, UKM Bangi, 43600, Selangor, Malaysia

[2]Department of Mechanical Engineering
Lotfy El-Sied st. off Gamal Abd El-Naser – Alexandria
Alexandria Governorate 11432, Egypt

[3]Faculty of Electrical Engineering, Universiti Teknologi MARA, 40450 Shah Alam, Selangor, Malaysia

*Correspondence: zambri@ukm.edu.my

Abstract. Investment and operating costs are the biggest obstacles in modernizing fish ponds in an otherwise very lucrative industry i.e. food production, in this region. Small-scale farmers running on small ponds could not afford to hire workers to man daily operations which usually consists of monitoring water levels, temperature and feeding fish. Bigger scale enterprises usually have some kinds of automation for water monitoring and replacement. These entities have to consider employing pH and dissolved oxygen (DO) sensors to ensure the health and growth of fish, sooner or later as their farms grow. This project identifies one of the sites, located in Malacca. In this project, water, temperature, pH and DO levels are measured and integrated with aerating and water supply pumps using Arduino. User could receive information at pre-determined intervals on preferred communication or display gadgets as long as they have internet. Since integrating devices are comparatively not expensive; it usually consists of Arduino board, internet and relay frames and display system, farmer could source these components easily. A sample of two days measurements of temperature, pH and DO levels show that this farm has a high-quality water. Oxygen levels increases in the day as sunshine supports photosynthesis in the pond. With this integration system, farmer need not hire worker at their site, consequently drive down operating costs and improve efficiency.

1. Introduction

1.1. World's aquaculture fish demand

Fisheries aquaculture remain important sources of food, nutrition, income and livelihoods for hundreds of millions of people around the world. According to The State of World Fisheries and Aquaculture 2016 [1], per capita fish supply reached a new record high of 20 kg in 2014, thanks to vigorous growth in aquaculture, which now provides half of all fish for human consumption, and to a slight improvement in the state of certain fish stocks due to improved fisheries management. In the last two decades, dramatic growth in aquaculture production has boosted average consumption of fish and fishery products at the global level. The shift towards relatively greater consumption of farmed species compared with wild fish reached a milestone in 2014, when the farmed sector's contribution to the supply of fish for human consumption surpassed that of wild-caught fish for the first time [1]. Total world fishery production (capture plus aquaculture) is projected to expand over the period, reaching 196 million tonnes in 2025. This represents an increase of 17% between the base period (average 2013–15) and 2025, but indicates a slower annual growth compared with the previous decade (1.5% versus 2.5%). Almost all of the increase in production will originate from developing countries. Their share in total output will increase from 83% in the base period to 85% in 2025. A more marked expansion is expected in Asia, with its share in total production rising from 70% to 73%. Surging demand for fish and fishery products will mainly be met by growth in supply from aquaculture production, which is expected to reach 102 million tonnes by 2025, 39% higher than the base period level.

1.2. Malaysia demand and production

Malaysia fisheries growths have increased steadily since three decades ago. In 1985 to 1997, Malaysia register growth of 4.5%. Most of the production was contributed by marine capture [2]. In 1997 the value added by the fisheries sector was estimated at about RM2 million representing about 1.5% of gross domestic product (GDP) or 11.1% of the agricultural GDP. Fisheries net export earnings were negative (-RM39.6 million) because Malaysia was a net importer of fish in terms of both quantity and value in 1997 [2]. Over the years, Malaysia rapid modernization on fisheries have improved fish products. Starting 2010, Malaysia has reached self sufficiency level (SSL) for most type of fish and therefore a net fish exporter since then on [3]. Figure 1 shows that Malaysia has achieved SSL level for 2010-2013(expected).

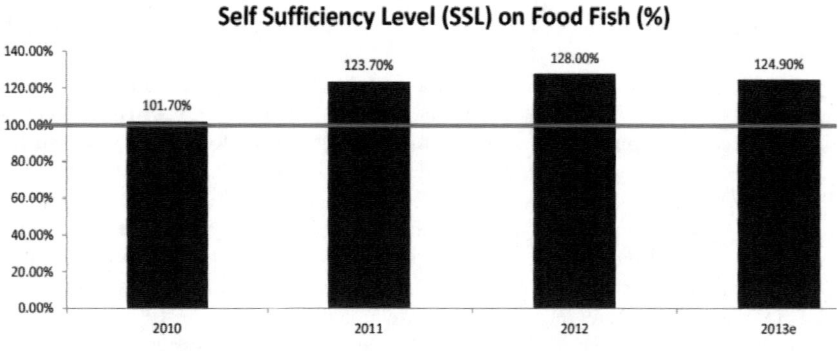

Figure 1: Malaysia's Self Sufficiency Level (SSL) on Fish Food (Source: Mazuki, H., 2015 [3].)

Moving further, the Malaysian government has realized the potential in aquaculture industry. Under the National Key Economic Areas (NKEA) for the 10th Malaysia ambitious, well-crafted development plan, the government has set up the Integrated Zone for Aquaculture (IZAQ) as one of the Entry Point Projects (EPP) to promote the industry, this is EPP6. The other two EPPs are the Integrated Cage Farming mainly for grouper, sea bass and tilapia and Seaweed Mini Estates. Figure 2 shows the projection under the IZAQ project. The 10 projects in IZAQ shall boost the growth to compound annual growth rate (CAGR) [3].

To ensure that projects rolled out are successful, the companies selected to trust into making the targets achieved, are companies with strong records and with established farming systems. For example, one of the anchor companies selected to achieve target 'an aquaculture park with large area dedicated to the organized production of high quality, fully certified shrimp for the premium market', is Hannan

Corp. Sdn Bhd. [4,5]. The big projects by Hannan Corp. involves farming areas more than 100 hectares, workers more than 400 and tasked to produce gross national income of RM500 million. The company has established its aquaculture expertise. This involves automation system, checking quality of water, pH, oxygen levels and etc. In this article, we propose a new approach to integrate devices usually used by aquaculture industries such pH, oxygen level and temperature sensors, pumps, aerators, lights and etc.

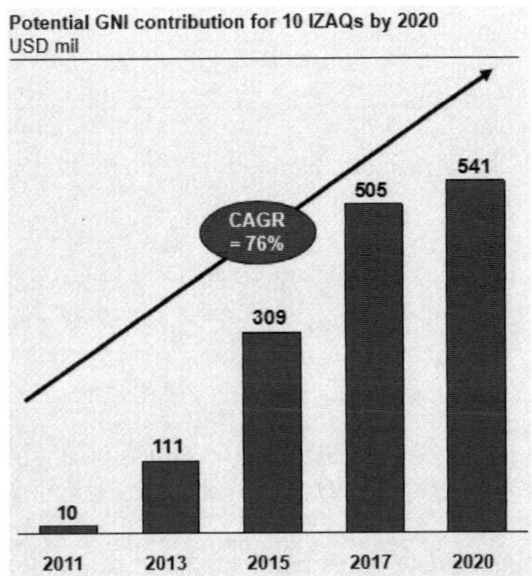

Figure 2: Malaysia's projection of the contribution of gross national income (GNI) for Integrated Zone for Aquaculture (IZAQ) is based on growth of 76%. (Source: Fisheries Department of Malaysia [3]).

2. Methodology

We have identified a site in Malacca as a typical aquaculture farm in Malaysia. The existing pond shown in Figure 3 has a length of 24 m, width of 13 m and depth of 1.2 m (typical water level). The pond produces tilapia and two of local Malaysian catfish favourites called *patin* and *keli*. The pond is located about 30 m from a creek, however the water level of pond is 8 m above the water level of creek on a typical day. There is also an abandoned well used to be a source of water supply using a suction pump.

Figure 3: Existing fish pond in Malacca, before real time fish monitoring and automation project started.

2.1. Instrumentations

Four sensors have been utilized in the proposed system; a temperature sensor to measure atmospheric temperature, a temperature sensor to measure water temperature, dissolved oxygen (DO) sensor and pH sensor. The specifications of these sensors are listed in Table 1.

Table 1: Specifications of the sensors implemented in the monitoring system.

	Model	Specifications
Temperature sensor	Digital Temperature Sensor (Dallas) – DS18B20	Range: -55 °C to +125 °C Accuracy: ±0.5 °C (-10 °C – +85°C)
pH sensor	Analog pH Meter Kit SKU: SEN0169	Measuring Range: 0-14 pH Measuring Temperature: 0-60 °C Accuracy: ± 0.1 pH (25 °C) Response Time: ≤ 1min
DO sensor	Atlas Scientific	Range: 0-35 mg/L Max Temperature: 50 °C Max Pressure: 690 kPa Max Depth 60 M Response time = 0.3 mg/L per second

The temperature sensor used here comes with stainless-steel-covered tip, made by Digital Temperature Sensor (Dallas) – model DS18B20. It can measure extreme temperatures -55 °C to 125 °C with very good accuracy.

A pH meter is a scientific instrument that measures the hydrogen-ion concentration (or pH) in a solution, indicating its acidity or alkalinity. The term pH was originally derived from a French word, 'pouvoir hydrogéne', which means 'hydrogen power'. This parameter shows the quantity of hydrogen ions (H^+) in the water. The scale for measuring the degree of acidity is called the pH scale, which ranges from 1 to 14. At 25 °C, a pH of 7.0 is considered neutral, while values below 7.0 are considered acidic, and above 7.0 are alkaline. The pH sensor used here, analog pH Meter Kit SKU: SEN0169, has the capability to measure the entire range of acidity to alkalinity (0-14). It can work continuously in turbid, strongly acid and alkali solution for months.

The Atlas Scientific DO sensor used here is founded on dissolved oxygen expressed in mg/L (weight of oxygen over volume of water) or its equivalent, parts per million (ppm). It can measure a wide range of oxygen 0-35 mg/L. Generally, all countries have an average of DO levels ranging between 5.0 and 9.0 in freshwater, and 6.5 and 9.0 for marine, all of which are within the limits of optimum fish production [6].

Table 2: Acceptable conditions for fish production in selected countries (PHILMINAQ, 2008 [6], National Water Quality Standards for Malaysia [7])

Parameters	Malaysia	The Philippines	Australia	General desired condition for freshwater pond
Temperature (°C)	*Normal + 2 °C	-	-	Depending on fish
pH (index 0-14)	6.5 - 9.0	6.5 - 8.5	5.0 - 9.0	6.5 - 9.0 *5 – 9
DO (mg/L)	3.0 – 7.0 *3 - 5	5.0	> 6.0	**Minimum DO that supports a large population of various fishes is 4-5.
Alkaline (mg/L)		-	> 20	5-500
NO_2-N (mg/L)	0.40	-	0.10	*** < 1
Phosphorus (P) in different forms	0.10 - 0.20 (P)	0.05 - 0.10 (P) in lakes and reservoir	< 0.10 (PO4)	Excess concentrations especially in rivers and lakes can result to algal blooms. It limits nutrient needed for the growth of all plants-aquatic plants and algae alike.

* Based on National Water Quality Standards for Malaysia for Fishery III - Common economic value and tolerant species and livestock drinking [7].
** Based on Water Research Center [15]
*** Based AO recommendation [16].

Solar radiation is the source of heat for all water bodies. The amount and angle of incidence of sunlight decides the energy entering the water body. Distribution of heat within the water by conduction is negligible because of very low heat conductivity of water especially in the tropics. Much of heat mixing takes place by convection aided by wind action. Heat is lost due to evaporation and also by direct exchange to air and substratum (the soil underneath) [8]. In artificial ponds, where the depth is usually 1 – 2 meters, there is only a minor temperature difference between the surface and water bed. The surface water temperature could be much higher in the afternoons and it is likely that fish take shelter in the cooler and deeper portions of the pond for comfort and probably to survive. Notwithstanding, fish mortality due to temperature change is a rare phenomenon in the tropical plains. It quite often fish mortality observed is not due to temperature, but a combination of factors e.g. temperature, low oxygen, metabolite load and salinity [8]. The temperature in Malaysia has very little fluctuation over the year. For example, an on-site measurement for three weeks in a weather station shows that temperature fluctuates +/- 1.5 °C from its mean [9].

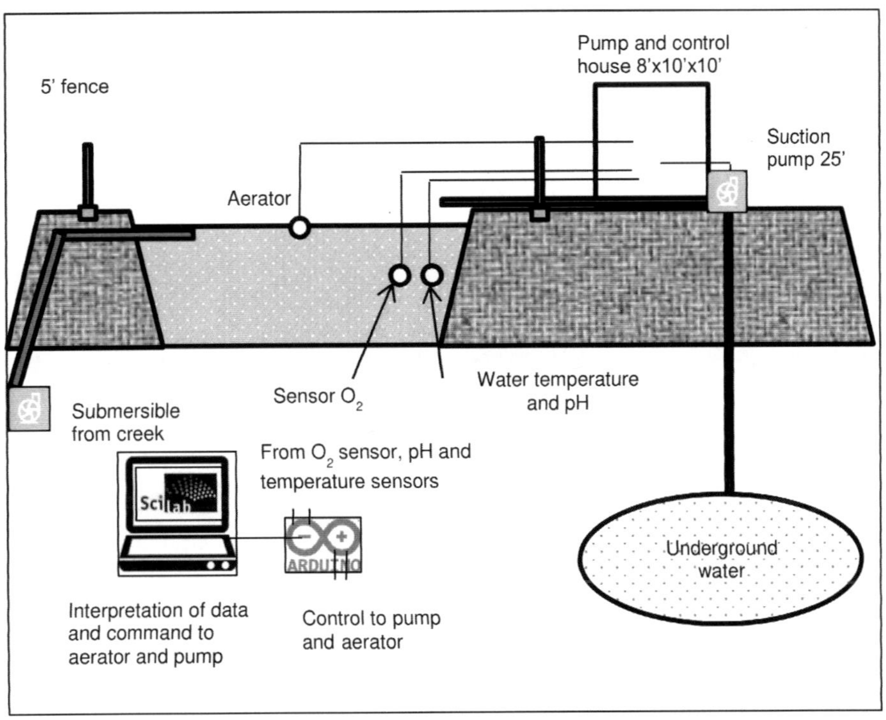

Figure 4: Schematic diagram of real time fish monitoring and automation project

The pH levels for natural waters range between 5.0 - 10.0 [6]. Based on the data shown in Table 2, the pH levels from three different countries i.e. Malaysia, The Philippines and Australia, fall within this range. The amount of oxygen consumption varies, depending on the size, feeding rate, activity level and species. Physical condition such as temperature, altitude, atmospheric pressures and salinity can also affect oxygen level. As temperature and pressure increase water holds less oxygen [6,12]. Therefore, for a typical fish pond system, a temperature sensor is necessary so that the reasons in the increase/decrease in chemical content could be traced, and might be attributed simply because of the rise in temperature rather than a direct leak (or other reasons) of these chemicals.

In aquaculture, alkalinity is the measure of the capacity of water to neutralize or buffer acids using carbonate, bicarbonate ions, and in rare cases, by hydroxide, thus protecting the organisms from major fluctuations in pH. Without a buffering system, free carbon dioxide will form large amounts of a weak acid (carbonic acid) that may potentially decrease the night-time pH level to 4.5. During peak periods of photosynthesis, most of the free carbon dioxide will be consumed by the phytoplankton and, as a result, drive the pH levels above 10.0. As discussed, fish grow within a narrow range of pH values and either of the above extremes will be lethal to them [6].

Other parameters for a fish pond operation are biological parameters Ammonia-Nitrogen (NH_3-N), Nitrate-Nitrogen (NO_3-N), Nitrite-Nitrogen (NO_2-N) and Phosphorous (P). There are also other parameters such as total solid (suspended or dissolved), heavy metals (mercury and lead) and pesticides. Monitoring the compliance and physical monitoring were identified as two critical factors that can ensure the success or failure of the monitoring system in place, however these are almost impossible for the at least for the Philippines [6].

In our proposed system, we have combined the most important parameters which are the temperature, pH and DO levels. The system as shown in Figure 4 also contains an aerator, a submersible pump (to pump water from the creek) and a suction pump (to extract water from the well). The Arduino codes for the temperature, DO and pH sensors were downloaded from the manufacturers' websites [10-12], respectively. The DO sensor, pH sensor and one of the temperature sensors were packed together and dipped in the water 1.5 m away from one side of the pond. The second temperature sensor was left in the atmosphere, just outside of the water. The algorithm for the Arduino was first loaded from a laptop, after that the Arduino board takes the data from the sensors directly and the instructions to aerator and pumps.

2.2. Data acquisition

The readings of the sensors were collected via an Arduino® Mega 2560 microcontroller. The connections of the different sensors to the Arduino board are illustrated in Figure 5 (a-c). The data was then sent via the existing, domestic Wi-Fi network to an online Google® spreadsheet. To send data via Wi-Fi, an extra Arduino shield (Cytron ESP8266) was fitted to the main board. The data passes through an [Arduino – PushingBox – Google Form – Google spreadsheet] chain. The detailed steps of the web connection setup can be found on [12]. Data from all sensors were captured and uploaded online at a rate of one set per minute. The Google spreadsheet was shared between the research team and farm owner.

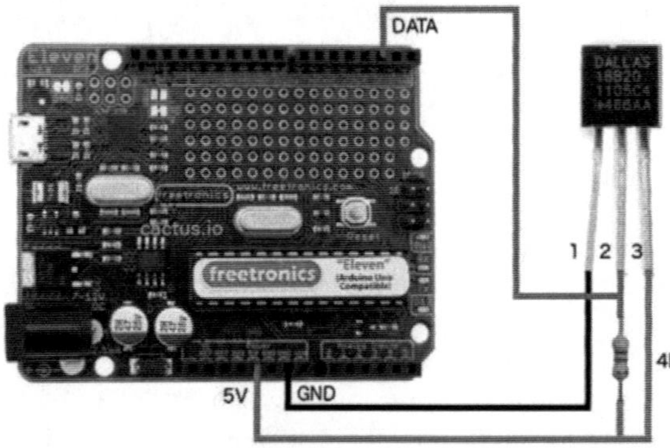

(a) Connections for temperature sensor

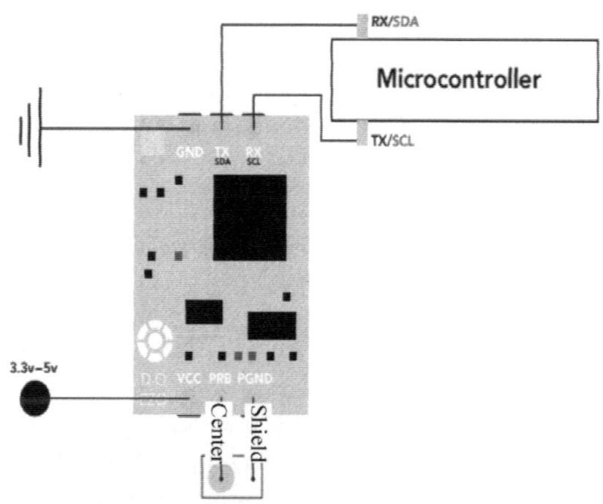

(b) Connections for DO sensor

IOP Conf. Series: Materials Science and Engineering **340** (2018) 012014 doi:10.1088/1757-899X/340/1/012014

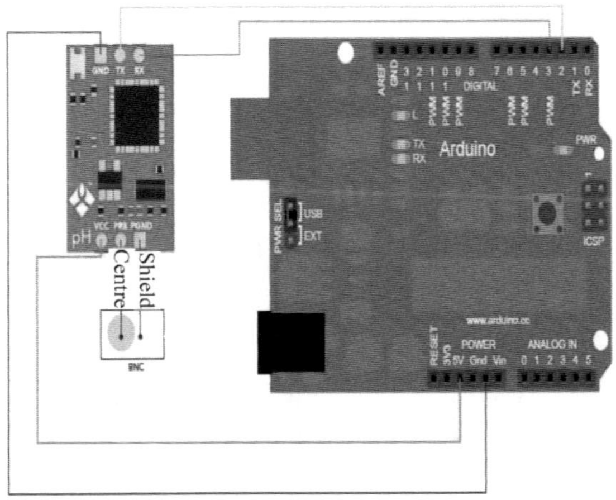

(c) Connections for pH sensor.
Figure 5: Connections from the different sensors to the Arduino board.

Beside the Google spreadsheet, data were monitored in-situ via a 16×2 LCD. The connections of the LCD to the Arduino are illustrated in Figure 6. Again, the Arduino code of the Wi-Fi shield was obtained from the manufacturer's website [14], whereas the LCD code is built-in the Arduino compiler. A display like this is necessary because farmers usually depend on wall-mounted display rather than an online system.

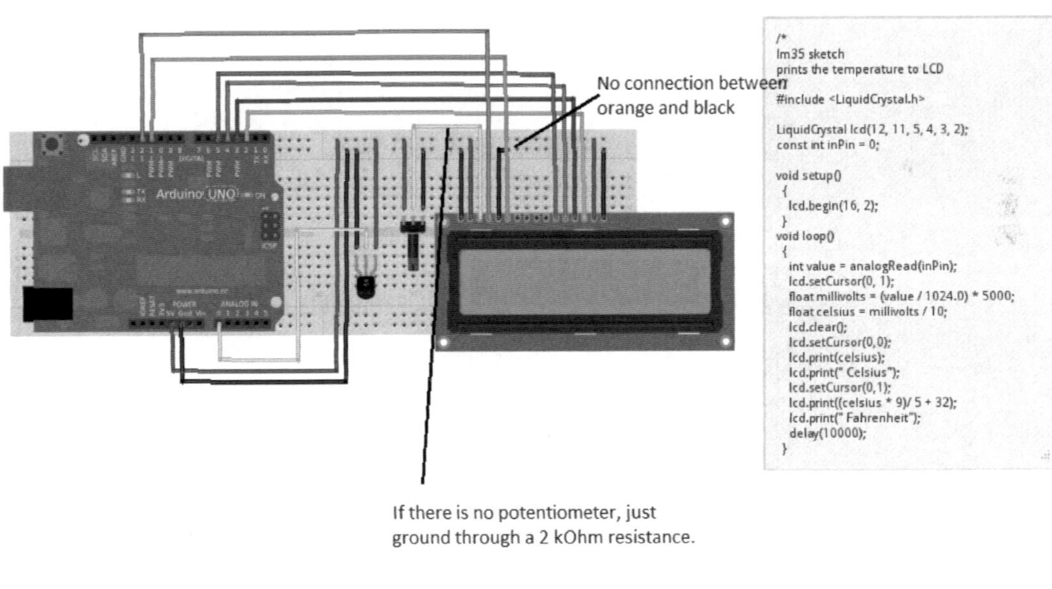

Figure 6: Connections of the LCD to the Arduino board.

3. Results and Conclusions

The researcher team met the farmer a few times to design an appropriate system and based on the conditions and requirement. A proper pump house was built where most of the devices are stored. An underground crossing from the pond to the creek was also built. This underground crossing is important because it houses a few pipes for water supply from the creek, excess water discharge back to the creek and electrical supply to the submersible pump. The pump house and the manhole serving the underground pipe is shown in Figure 7.

(a) Pump house (b) Manhole serving underground pipes

Figure 7: Physical installation of the fish automation and monitoring

Figure 8 shows actual acrylic container which houses the Arduino board, Wi-Fi shield, relay, sensor boards and the display. To the left, there is a fan, to ensure the temperature does not go out of control within the container. The first and second readings (top line) are atmospheric and water temperature i.e. T_a = 27.7 °C and T_w=29.7 °C. This picture was taken at about 6pm, where the ambient temperature has dropped but the water temperature is still high. The bottom line shows DO and pH level at 9.52PPM and 6.03.

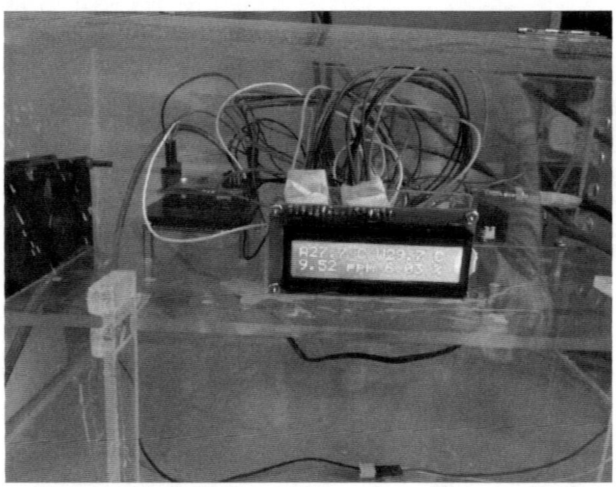

Figure 8: Actual installation and sample reading on display

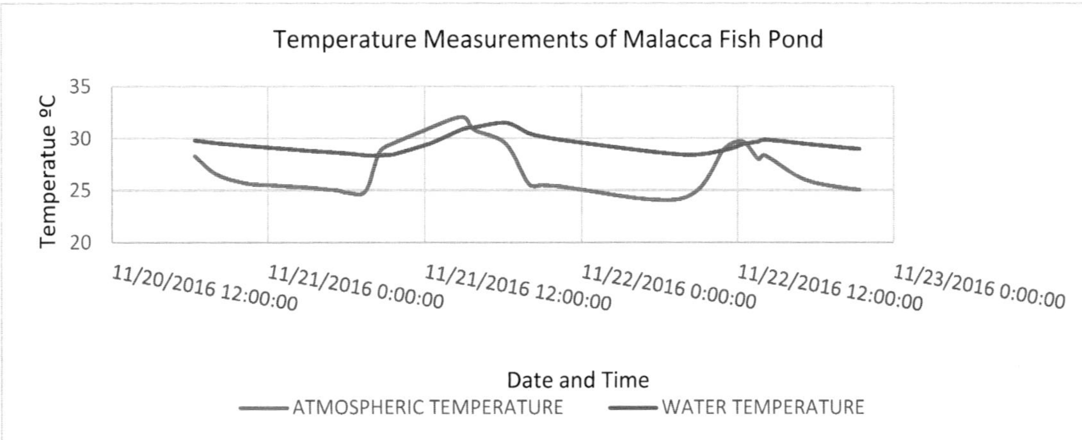

Figure 9: Temperature Measurements of Malacca Fish Pond

Figure 9 shows the water temperature and ambient temperature. This data was collected in November 2016, where the North East Monsoon is about to start. The water temperature is high with a mean of approximately 30 °C. The ambient temperature fluctuated bigger with a maximum +/- 3 °C from its mean, slightly bigger but similar to [8]. It seems that the pond temperature responded to ambient temperature a slow pace with a small lag.

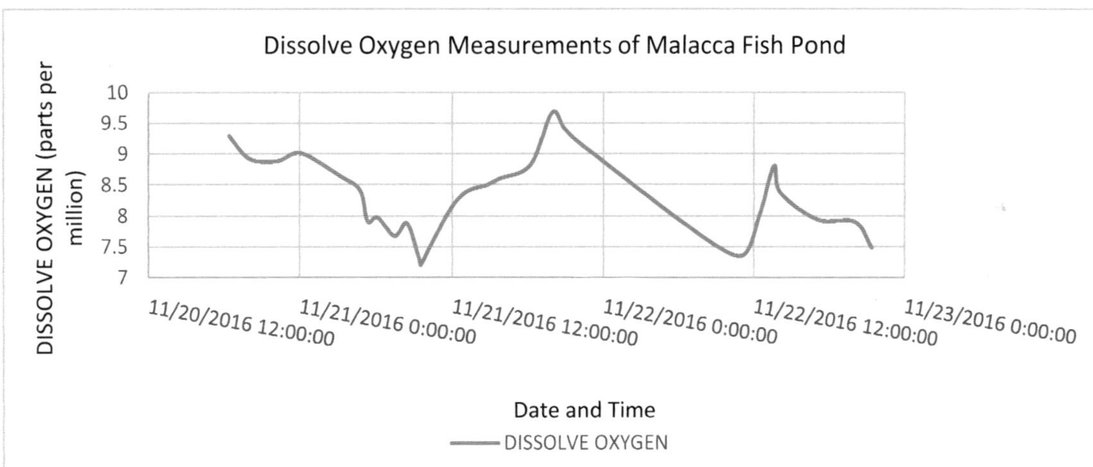

Figure 10: Dissolve Oxygen (DO) Measurements of Malacca Fish Pond

Figure 10 shows the DO levels for the two-day measurements. The fluctuation is relatively high 7-9.5 mg/L or PPM. The DO levels follow the sun cycle, confirming previous study [6] where pond produces oxygen via photosynthesis by phytoplankton, when there is direct sunlight.

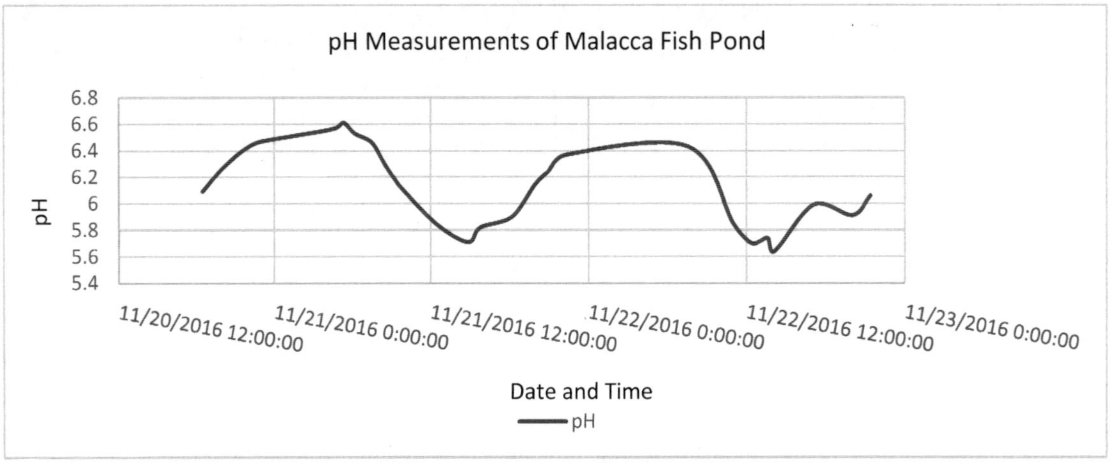

Figure 11: pH Measurements of Malacca Fish Pond

Figure 11 shows the pH measurement for the 2 days. It follows the daily cycle, the photosynthesis process by phytoplankton consume carbon dioxide (CO_2), driving down the pH level at noon. Figure 9-11 show that these important parameters are well within the good conditions for aquaculture shown in Table 3. The material costs for the Arduino boards, circuits, Wi-fi, relay are within RM1000, most of the costs went to construction of infrastructure and pumps. With the very minimum costs like this, farmers are able to monitor and automate their pond more efficiently.

This project has become a reference for ongoing capstone/design projects and final year projects (FYP) for the authors. The industry-extensive involvements especially both at the design/requirement and installation stages provide students a lot of experience dealing with industrial requirements. The involvements of researchers in the industry such as this will help the long-term aim of the country towards modernization [17].

Acknowledgements

The researchers would like to express gratitude for the Public-Private Research Network (PPRN) grant 100/IRMI/GOV/ 16/6/2, owner Mr. Omar Md. Diah and the Ministry of Education for providing the Exploratory Research grant ERGS/1/2013/TK01/UKM/03/2.

References

[1] The State of World Fisheries and Aquaculture (2016). Food and Agriculture Organization of the United Nations.
[2] Ahmad, A. T., Salim, K., Ean, C. P., Isa, M. M., & Fong, L. C. (2003). An overview of the socio-economic status of fisheries in Malaysia. Fisheries Research Institute, 11960.
[3] Mazuki, H. (2015). Industry and Market Status of Tilapia in Malaysia, 4th International Trade and Technical Conference And Exposition on Tilapia (Tilapia 2015).
[4] Yusof M., Presentation, Program National Key Economic Areas (NKEA) on Agriculture, Fisheries Department of Malaysia, Downloaded on 15 Feb. 2017. http://etp.pemandu.gov.my/upload/NKEA_Agriculture.pdf
[5] ETP Annual Report (2014) on Agriculture, The Ministry of Agriculture and Agro-Based Industry, Downloaded on 21 Feb. 2017. http://etp.pemandu.gov.my/annualreport2014/upload/13_NKEA11_ENG_Agriculture.pdf
[6] PHILMINAQ (Mitigating impact from aquaculture in the Phillippines), 2008. Water Quality Criteria and Standards for Freshwater and Marine Aquaculture, Marine Science Institute, Univ. of the Philippines, downloaded on 21 Feb. 2017,

http://aquaculture.asia/files/PMNQ%20WQ%20standard%202.pdf
[7] National Water Quality Standards For Malaysia (2008), downloaded on 21 Feb. 2017, http://www.wepa-db.net/policies/law/malaysia/eq_surface.htm
[8] Kutty, M. N. (1987). Site selection for aquaculture: physical features of water, Nigerian Institute for Oceanography and Marine Research, Food and Agriclture Organization of The United Nations. Downloaded on 21 Feb. 2017, http://www.fao.org/documents/en/detail/69895/
[9] Harun Z., Reda, E. and Zulkifli, R. (2017). Buoyancy effect on atmospheric surface layer: measurements from the East Coast of Malaysia, IOP Conference Series.
[10] https://www.milesburton.com/Dallas_Temperature_Control_Library
[11] https://www.dfrobot.com/wiki/index.php/Analog_pH_Meter_Pro_SKU:SEN0169
[12] http://www.atlas-scientific.com/product_pages/circuits/ezo_do.html
[13] http://www.instructables.com/id/Post-to-Google-Docs-with-Arduino/
[14] http://cytron.com.my/p-shield-esp-wifi
[15] Water Research Center website, downloaded 21 Feb. 2017, http://www.water-research.net/index.php/dissovled-oxygen-in-water
[16] Somerville, C., Cohen, M., Pantanella, E., Stankus, A. & Lovatelli, A. (2014). Small-scale aquaponic food production. Integrated fish and plant farming. FAO Fisheries and Aquaculture Technical Paper No. 589. Rome, FAO.
[17] Harun, Z., Khamis, N. K., Isa, M. D., Mohamed, Z. and Hashim, H. (2013). The roles of professional engineers at the institutions of higher learning in nation-building, International Education Studies, vol. 6(6), 137.

IOP Conf. Series: Materials Science and Engineering **340** (2018) 012015 doi:10.1088/1757-899X/340/1/012015

Application of Particle Swarm Optimization Algorithm for Optimizing ANN Model in Recognizing Ripeness of Citrus

Anis Diyana Rosli[1], Nur Sabrina Adenan[2], Hadzli Hashim[2], Noor Ezan Abdullah[2], Suhaimi Sulaiman[3], Rohaiza Baharudin[1]

[1]Faculty of Electrical Engineering
Universiti Teknologi MARA Malaysia – P.Pinang

[2]Faculty of Electrical Engineering
Universiti Teknologi MARA Malaysia – Selangor

[3]Faculty of Electrical Engineering
Universiti Teknologi MARA Malaysia – Terengganu

Abstract. This paper shows findings of the application of Particle Swarm Optimization (PSO) algorithm in optimizing an Artificial Neural Network that could categorize between ripeness and unripeness stage of citrus suhuensis. The algorithm would adjust the network connections weights and adapt its values during training for best results at the output. Initially, citrus suhuensis fruit's skin is measured using optically non-destructive method via spectrometer. The spectrometer would transmit VIS (visible spectrum) photonic light radiation to the surface (skin of citrus) of the sample. The reflected light from the sample's surface would be received and measured by the same spectrometer in terms of reflectance percentage based on VIS range. These measured data are used to train and test the best optimized ANN model. The accuracy is based on receiver operating characteristic (ROC) performance. The result outcomes from this investigation have shown that the achieved accuracy for the optimized is 70.5% with a sensitivity and specificity of 60.1% and 80.0% respectively.

1. Introduction

In Malaysia, the agriculture sector is growing and increasing demand among consumers. One of them is "Limau Madu" or scientific name is citrus suhuiensis. It belongs of the rutaceace family and one of the local fruit that has high potential for development and demand. It has a sweet taste, content of juice is high, good fruit size and have less fiber and bagasse [1]. Nowadays, fruit industry is important in term of economic and social. There are a variety of local fruits available either seasonal or non-seasonal such as watermelon, citrus, pineapple, banana and etc. Based on new technology and research done, fruits can be cloned so that the production of fruit can be increased and quality of fruit can be control [2]. However, the research on how to identify maturity of fruit are still ongoing progressively and widely available. Traditional way of identifying ripeness of fruit can be determined by an expert person that has experience and skill. Recently, many researches were done in recognizing ripeness of citrus by using PSO algorithm and Artificial Neural Network [4-6].

The ability to classify the quality either ripeness or unripeness will benefit for both farmers and consumers. Besides that, it will help the farmer to grade the fruit in different size and price. To commercialize a local fruit such as citrus suhuiensis hort., many aspects must be considered in order to increase its productivity. It was reported that the process of maturity stage is within the 8-10 month, where the skin will become oily glands and soft [3]. However from observations, the skin color is the same whether it is within growth stage or maturity stage.

The objective of this work is to recognize ripeness and unripeness of Citrus Suhuensis using Particle Swarm Optimization algorithm by adjusted their connections weights with different hidden layer size. The new weights are predicted by PSO algorithm during training of an Artificial Neural Network (ANN). A reflectance spectrometer is used as a tool for nondestructive determination of the citrus's ripeness condition. The percentage reflectance from the spectrometer is based on wavelength of green, orange and yellow color.

2. Instrumentation

2.1. *Visible Spectrum (VIS)*

By viewing a substance through a spectrometer, one can distinguish the exact mixture of color, which corresponds to specific wavelengths of light. VIS has been chosen as the light reflectance, where it consists of a spectrum of wavelengths which range from of 400 to 780nm as shown in Figure 1 [7]. This visual spectrum is used in order to see through the sample experiments and it can be related to the electrical instrument (spectrometer) being used. These wavelengths are used by spectrometer in order to collect data in term of percentage and reflected light that received back to the spectrometer.

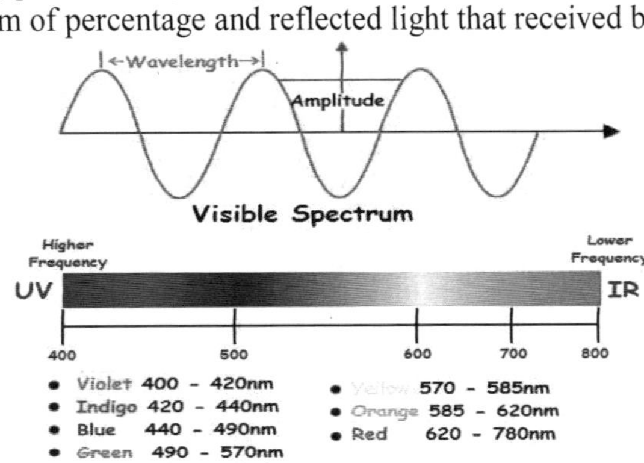

Figure 1: Optical properties of VIS

IOP Conf. Series: Materials Science and Engineering **340** (2018) 012015 doi:10.1088/1757-899X/340/1/012015

2.2. Spectrometer

A spectrometer is a device that splits light into various colors. The waves of visible light can see as color of rainbow, where each color has different wavelengths. In this project, spectrometer is used to measure the amount of light. Spectrometer is a device that transmit light beam from a source and read from intensity of light reached by receiver or detector as shown in Figure 2. It provides optical measurements in the ultra-violet (UV)/visible spectrum (VIS) and Near Infra Red (NIR) spectral range. The spectrometer, MCS600 is suitable for measurements in the UV range, versatile tasks and a very high degree of flexibility can be fulfilled [8, 9, 10]. The spectrometer is attached with the head labelled as OFK30. This spectrometer can cover a spectra range between 190 to 2200 nm [11]. The result obtained from the light reflectance head is interpreted by Aspect Plus software.

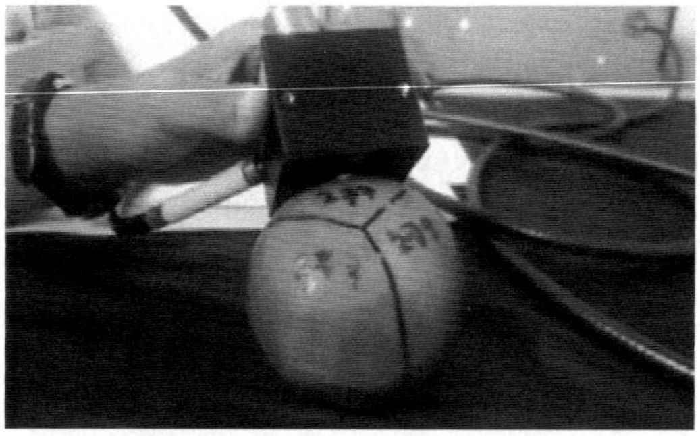

Figure 2: Spectrometer

2.3. Particle Swarm Optimization

Particle swarm optimization (PSO) is a new method of optimization developed in 1995 by Kennedy and Eberhart [12]. The idea of PSO was inspired by social behavior of bird flocking, fish schooling and as well as swarming theory. The concept of PSO is that, each particle will searched space to find the best solution (fitness) which called pbest. Every single particle has their own pbest value even though it can't achieve the global value called gbest. Each particle has the ability to memorize and track the current best particle (pbest) in a swarm [13]. In the proposed method, the solution vector represents all the ANN's weight values. In order to minimize the problem of ANN training to PSO algorithm, the weights and biases of a network are predicted by PSO algorithm.

Generally, PSO used to replace the architecture of the Artificial Neural Network (ANN) because the train in ANN takes a lot of time and it only based on trial and error [14]. The basic concept of PSO technique lies in accelerating each particle towards its pbest and the gbest locations at each time step. Acceleration has random weights for both its pbest and the gbest locations [15]. These are different between each other and greatly depend on hidden layer size, input, bias and output. Two factors characterize a particle status on the search space: its position and velocity. The most popular formulations of how particle adjusts its velocity and position are shown in equations (1) and (2).

$$VI(t+1)(d) = Wvit(d) + C1r1(Pbi(d) - xit(d)) + C2r2(Gbi(d) - xit(d)) \quad (1)$$
$$Xi(t+1)(d) = xit(d) + vi(t+1)(d) \quad (2)$$

Where, d is the indexes of dimension in the search space, represents the inertia weight, and are regarded as cognitive and social parameters for algorithm respectively, and are two random numbers is the personal best position recorded by particle I, while is the global best positions obtain by any particle in the population [16].

3. Methodology

3.1 Data Collection.

Data were taken from previous researcher. The samples of citrus were classified first by an expert person for either ripeness or unripeness. 200 samples of citrus were used for this project where 100 samples for each ripeness and unripeness respectively. Each sample's skin was divided into 6 portions, and data reflectance by spectrometer was measured for each of this portion. Thus, the total data for ripeness and unripeness is 1200 and 600 respectively. Table 1 shows an example of 10 samples of the required data being measured.

Table 1: Sample of Data

No.	**Wavelength**	**Reflectance (%)**	
		Ripe	**Unripe**
1	509.144	3.57143	6.47149
2	512.514	4.35931	8.32117
3	515.884	5.93434	8.40336
4	519.255	7.07317	15.0769
5	569.819	15.8774	14.1892
6	573.190	14.8501	13.8085
7	576.561	13.9893	13.0147
8	590.043	11.4418	11.5526
9	593.414	10.6809	11.4328
10	596.784	10.5017	11.0696
OUTPUT		**1**	**0**

Table 2 shows the proposed training and testing data set applied in this experiment where the classification output is set as a '0' for unripeness and '1' for ripeness respectively. These data were used to design and train for optimizing 6 ANN model connections weights using PSO algorithm with various hidden layer size. All these models were then validated with the testing set. As depicted in the table, the ratio between training and testing data sets are 400:400 and 100:100 respectively.

Table 2: Sample of Data of Proposed Set

Classes	**Training**	**Testing**	**Output**
Ripeness	400	100	1
Unripeness	400	100	0

3.2 Artificial Neural Network

The main goal of artificial neural network modeling is to obtain the weights and minimize the error. The weights of NN training can adjust by default or another algorithm. ANNs are a problem-solving tool that has become an alternative modeling method to some physical and nonphysical systems with scientific or mathematical basis [14]. It nearly attitude of the process of human learning and this intelligent are train like human thinking. Just as human brains can be trained to master some tasks through experiential knowledge and training, ANNs can be trained to recognize patterns and perform optimization through a training process.

The architecture of neural network and PSO are shown in Figure 3. PSO were predicted for each weight and the values were used to train into ANN. For every single number of neuron in the hidden layer, it provides various connections and thus would affect the parameter called as dimension. These dimensions were use as parameter to implement into PSO technique. Therefore, these parameters need to be adjusted for optimization. The total connections (or parameters) plus the necessary bias connections can be computed as below [17].

$$\text{No. of connection s} = (\text{input} \times \text{hidden}) + \text{bias}_{\text{hidden}} + (\text{hidden} \times \text{output}) + \text{bias}_{\text{output}} \quad (3)$$

Based on data training set described in Table 2, the best optimized model is with respect to hidden layer size. These 6 designed models were later being validated using the testing data set. Classification accuracy will be explained in the next section.

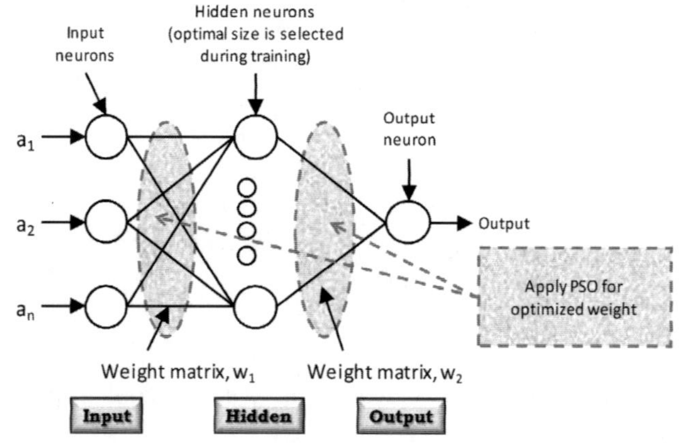

Figure 3: A typical architecture feed forward of a neural network with PSO training

3.3 Model's Accuracy

The model's accuracy could be described in terms of confusion matrix. From the confusion matrix, the best accuracy, sensitivity and specificity could be computed. Accuracy is calculated based on value of true positive, true negative, false positive and true negative. Besides that, the threshold value also influence the percentage accuracy, and it could be observed on the Receiver Operating Curve (ROC) plot. Optimization of the trained models was decided using a confusion matrix [18]. Confusion matrix is a matrix for a two-class classifier, contains information about actual and predicted classifications done by a classification system [19]. Table 3 describes the confusion matrix for a two cases classifier.

Table 3: Confusion Matrix

		Actual/Target	
		True (-ve)	True (+ve)
Predicted Class	Predict (-ve)	TN	FN
	Predict (+ve)	FP	TP

The entries in the confusion matrix have the following meaning is defined as:

- Predicted (+ ve) - the correct predictions
- Predicted (- ne) - the incorrect predictions
- *A*ctual (+ ve) - the actual correct number
- Actual (- ne) - the actual incorrect number
- *TN* (true negative) - the number of correct predictions that an instance is negative
- *TP (*true positive) – the number of correct predictions that an instance is positive
- *FP* (false positive) - the number of incorrect predictions that an instance is positive
- *FN* (false negative) - the number of incorrect of predictions that an instance negative

Sensitivity and specificity are commonly used terms that describe the accuracy of a test. Sensitivity is a measure of the ratio or percentage of true positive (*TP*) and a positive actual test (*A*+). It is also known as true positive rate (*TPR*) which is the proportion of positive cases that were correctly identified, as calculated using the equation:

$$\text{TPR} = \frac{\text{TP}}{\text{TP} + \text{FN}} \tag{4}$$

Specificity is a measure of the ratio or percentage of true negative (*TN*) and a negative diagnosis test (*A*-). It is also called as true negative rate (*TNR*) which is the proportion of negative cases that were correctly identified, as calculated using the equation:

$$T\text{NR} = \frac{\text{TN}}{\text{TN} + \text{FP}} \tag{5}$$

Percentage accuracy is calculated from the ratio of *TPR* plus *TNR* with the total of all samples. The equation is shown below:

$$\text{Accuracy (\%)} = \frac{\text{TP} + \text{TN}}{\text{TP} + \text{FN} + \text{TN} + \text{FP}} \times 100 \tag{6}$$

At the final satge of this investigation, the performances of all the trained ANN models were analyzed by observing the most appropriate threshold level which would be decided by analyzing the minimum *Euclidean Distance* (*ED*) values from the receiver operating characteristic (ROC) plot [20].

4. Result and Discussion

4.1 ANN Model Designing

The 6 models were designed of ANN which focuses on 5, 10, 15, 20, 25 and 30 neurons located in the hidden layer size of the network architecture. The total samples of data used were 1000 for both ripeness and unripeness citrus. These data were divided into two parts where 800 is for training set and 200 is for testing set. From the experiment done, the best optimized model has 20 neurons located in the hidden layer. This optimized model produced an accuracy of 70.5% as shown in Table 4. The best model would have the highest accuracy. From the table, this model has the highest accuracy and the identified best threshold is computed as 0.7 as shown in Figure 4.

Table 4: Accuracy and Respective Threshold of Various Hidden Layer Size

Neurons	Threshold	Accuracy(%
5	0.5	60.0
10	0.5	50.0
15	0.5	64.5
20	0.7	70.5
25	0.3	60.5
30	0.1	64.0

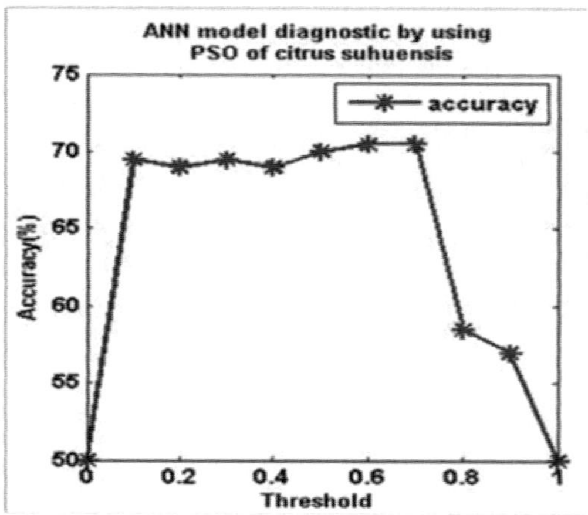

Figure 4: Accuracy ANN Model by using PSO of citrus.

Every hidden layer size has different accuracy when tested with the testing data set. The accuracy of each model was based on value of threshold from 0 to 1 from the ROC plot and confusion matrix. Table 5 shown the confusion matrix after the experimental work and it contains values for TN, TP, FN and FP.

Table 5: Confusion Matrix After Experiment

		Actual/Target	
		True (-ve)	True (+ve)
Predicted Class	Predict (-ve)	80	39
	Predict (+ve)	20	61

Besides that, the results also show in term of sensitivity and specificity. The values for sensitivity (TPR ripeness), specificity (FPR unripeness) and accuracy, could be calculated in manually. Based on equation (4), (5) and (6) the value of accuracy, TPR for ripeness and TNR for unripeness are calculated as below. Analytically, this model could recognize and classify better if tested with unripeness citrus rather than freshness where the specificity scores 80%.. However, its capability is reduced to 61% when recognizing ripeness citrus.

$$TPR = \frac{61}{61 + 39} \times 100 = 61.0\%$$

$$TNR = \frac{80}{80 + 20} \times 100 = 80.0\%$$

$$Accuracy\,(\%) = \frac{80+61}{80+39+20+61} \times 100 = 70.5\%$$

Receiver Operating Characteristics (ROC) usually is used to show the true positive rate (TPR) versus the false positive rate (FPR) across multiple thresholds. It is used to verify the optimal threshold and to maximize classification accuracy and minimize classification errors [19]. To get the best accuracy, it must consider the best value of threshold that is closer in terms of Euclidean distance to the ideal point (0,1).

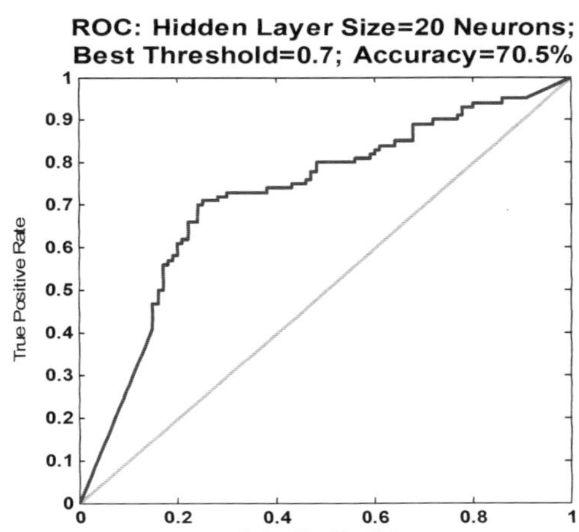

Figure 5: ROC of hidden layer size with 20 neurons

Figure 5 shows the ROC plot for hidden layer size with 20 neurons. From the figure, the best threshold is 0.7 where at that point the value for FPR is low at 0.2 while the TPR is the highest at 0.61. In addition, the curve also shows closeness, leaning towards to the y-axis and x-axis.

5. Conclusion

This paper presents the implementation of Particle Swarm Optimization (PSO) algorithm to artificial neural modeling in order to determine the ripeness and unripeness of citrus suhuensis. 200 citrus were being experimented from 1000 raw samples they were divided into ripeness and unripeness. The visible spectrometers were used from 445nm until 609nm based on yellow, green and orange color. From the range, only 30 identified wavelengths were selected for each sample's input for the experiment. There training set was used to train in designing the best optimized connections' weights using PSO. The algorithm has produced new weights during the training phase. When validated with the testing set, the best optimization model achievement is when the hidden layer size is 20 neurons. At this stage, the highest overall accuracy is recorded as 70.5% respectively based on 0.7 threshold. Also, the sensitivity and specificity is 61% and 80% respectively. It can be concluded that this model could recognized better a non-ripeness of citrus compared to ripeness.

6. Acknowledgement

The authors would like to give their appreciation to Faculty of Electrical Engineering, Universiti Teknologi MARA Shah Alam, Research Management Institute (RMI) and Kementerian Pengajian Tinggi for funding research work under RAGS grant code (600-RMI/RAGS 5/3 (27/2014)).

References

[1] F. N. Osman, "A Preliminary Study of Citrus Suhuensis Flavor Based on RGB Color Information," Thesis Bachelor of Engineering (Honours) in Electronics Engineering (EE241), 2011

[2] L.Huijun and W.Xiangfeng, "Rapid Shelf-life Identification Model of Citrus Based on Near Infrared Spectroscopy" *Knowledge Acquisition and Modeling, 2008, KAM '08 International Symposium on*, pp 298 – 301, 2008

[3] P. P. D. Manjung (2010, PANDUAN MENANAM LIMAU-Tanaman Jangka Panjang). Available: http://pertanianmjg.perak.gov.my/bahasa/panduan_tanamlimau.htm#penyakit

[4] H.Ji and J.Yuan. "The application study of apple color grading by particle swarm optimization neural networks." *Intelligent Control and Automation, 2006. WCICA 2006, The Sixth World Congress on*, Vol. 1. IEEE, pp 2651 – 2654, 2006

[5] H.Saad, A.P Ismail, N. Othman, M.H Jusoh, N.F Naim, N.A Ahmad, "Recognizing the ripeness of bananas using artificial neural network based on histogram approach." *Signal and Image Processing Applications (ICSIPA), 2009 IEEE International Conference on,* IEEE, pp 536 - 541 2009.

[6] M.Recce, J.Taylor, A.Piebe and G.Tropiano, "High speed vision-based quality grading of oranges" *Neural Networks for Identification, Control, Robotics, and Signal/Image Processing, 1996. Proceedings., International Workshop on*, pp 136 – 144, 1996

[7] Visible spectrum of wavelengths which range from of 400 to 780nm, 2014, Available: http://chemistry.tutorcircle. com/inorganic chemistry/visible-spectrum.html

[8] R.Riesenberg and A.Wuttig,,"Intelligent Photonic Measurement For Spectrophotometers" *XVII IMEKO World Congress Metrology in the 3rd Millennium*, pp 202-206, 2003

[9] D.H. Card, D.L. Peterson and P.A. Matson, "Prediction of leaf chemistry by the use of visible and near infrared reflectance spectroscopy," *Remote Sensing of Environment*, Vol. 26, Issue 2, pp 123–147, 1988

[10] T.A. Lestander, P.Geladi, S.H. Larsson, and M.Thyrel, "Near infrared image analysis for online identification and separation of wood chips with elevated levels of extractives", *Journal of Near Infrared Spectroscopy,* Volume 20 Issue 5, Pages 591–599, 2012

[11] C. Zeiss, "Multi Channel Spectrometer System MCS 600 Model MCS 621 VIS II/611 NIR 1.7/CLH 600/OFK 30 (manual book), Zeiss", 2008

[12] G.Qinqing, Z.Guanping, C.Dexin and H.Ketai, "Image Enhancement Technique Based On Improved PSO Algorithm", *Industrial Electronics and Applications (ICIEA), 2011 6th IEEE Conference on*, pp.234-238, 2011.

[13] S.M.W.Masra, P.K.Pang, M.S.Muhammad and K.Kipli," Application of Particle Swarm Optimization in Histogram Equalization for image Enhancement" *Humanities, Science and Engineering (CHUSER), 2012 IEEE Colloquium on*, pp 294 – 299, 2012

[14] Q.Abbas, J.Ahmad, and W.H.Bangyal. "Dynamic Hidden Layers Selection of ANN Architecture Using Particle Swarm Optimization." *IACSIT International Journal of Engineering and Technology*, Vol. 5, No. 2, pp 195-197, 2013

[15] V.G Gudise and G.K Venayagamoorthy, "Comparison of Particle Swarm Optimization and Backpropagation as Training Algorithms for Neural Networks", *Swarm Intelligence Symposium, 2003. SIS '03. Proceedings of the 2003 IEEE,* pp 110 – 117, 2013

[16] T. Gong and A. L. Tuson, "Particle Swarm Optimization for Quadratic Assignment Problem A Forma Analysis Approach", *International Journal of Computational Intelligence Research*, Vol. 4, No.2, pp 177-185, 2008.

[17] M. Hassoun,"Fundamentals of Artificial Neural Networks," MIT Press, Bradford Book, 1995

[18] P.Martiskainen, M.Jarvinen, J.P.kon, J.Tiirikainen, M.Kolehmainen, and J.Mononen, "Cow behaviour pattern recognition using a three-dimensional accelerometer and support vector machines," *Applied Animal Behaviour Science*, vol.119, pp.32-38,2009.

[19] T. Fawcett, "ROC Graphs: Notes and Practical Considerations for Researchers", Kluwer Academic Publishers, 2004.

[20] B.O'Connel and H.Myers, "Receiver Operating Characteristic (ROC) curves," *Journal of Clinical Nursing*, vol. 11, pp. 134-136, 2002

Embedded 32-bit Differential Pulse Voltammetry (DPV) Technique for 3-electrode Cell Sensing

Aqmar N.Z.N[1], W.F.H. Abdullah[1], Z.M.Zain[2], S.Rani[1]

[1]Faculty of Electrical Engineering, Universiti Teknologi MARA, Shah Alam Selangor
[2]Faculty of Applied Science, Universiti Teknologi MARA, Shah Alam Selangor

E-mail: zatieyl@gmail.com

Abstract This paper addresses the development of differential pulse voltammetry (DPV) embedded algorithm using an ARM cortex processor with new developed potentiostat circuit design for in-situ 3-electrode cell sensing. This project is mainly to design a low cost potentiostat for the researchers in laboratories. It is required to develop an embedded algorithm for analytical technique to be used with the designed potentiostat. DPV is one of the most familiar pulse technique method used with 3-electrode cell sensing in chemical studies. Experiment was conducted on 10mM solution of Ferricyanide using the designed potentiostat and the developed DPV algorithm. As a result, the device can generate an excitation signal of DPV from 0.4V to 1.2V and produced a peaked voltammogram with relatively small error compared to the commercial potentiostat; which is only 6.25% difference in peak potential reading. The design of potentiostat device and its DPV algorithm is verified.

Keywords - differential pulse voltammetry (DPV), potentiostat

1. Introduction

Over the past few decades, electrochemical devices have been developed and used broadly in industrial, biotechnology, physics and chemistry laboratories to help the researchers conduct their experiments . Researchers use these instruments to process the required sample collection and make analysis on the chemical reactions. Potentiostat is one of the most important hardware in electrochemical instruments. It is normally used with 3 kinds of electrode known as working electrode(WE), reference electrode(RE) and counter electrode(CE) as its sensing cell. Potentiostat is used to control the voltage difference between a working electrode and reference electrode in electrochemical cell. It also measures the current flow between the working electrode and counter electrode. Although this instrument is an important element in electrochemical research, they have seen relatively little application in resource poor settings, such as undergraduate laboratory courses and the developing world due to the cost factor. Hence, it is required to develop a device that can overcome this circumstances.

Most of the researchers focus on the analog design of the potentiostat [1],[2],[3], but very few discussing about the development of its software algorithm for 3-electrode cell sensing[4]. In this paper, the DPV-based embedded to allow in-situ 3-electrode cell sensing is discussed.

DPV is one of the frequently used voltammetry techniques [5],[6],[7] for detecting analytes in flow systems, including the in vivo analysis of a patient's blood and DNA. Furthermore, DPV is suitable for the rapid analysis of a single analyte. In DPV, the current is sampled twice in each pulse period and the difference between these two current values is recorded to produce a peaked voltammogram. The height peak is directly proportional to the concentration of analyte. DPV has benefit over the other analytical technique as it can reduce the effect of capacitive current and improves the signal-to-noise ratio by attenuating the background current[8]. Figure 1 shows the differential pulse voltammetry and its produced peaked voltammogram.

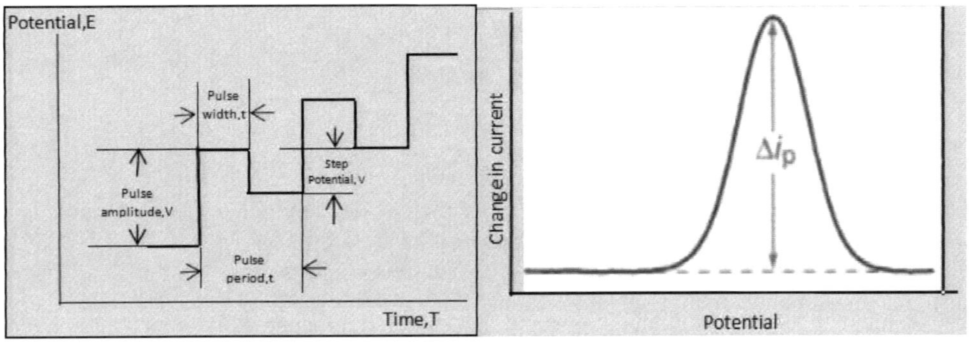

Figure 1 DPV excitation signal potential and its voltammogram

2. Materials and Method

2.1 Experiment Setup

The developed system is composed of an MCU component LPCI768, potentiostat and 3-electrode system. The LPCI768 is a 32- bit ARM Cortex-M3 based MCU for embedded applications. It is used to program the excitation signal voltage for potentiostat, measure and converting the analog to digital signal of the chemical response current and store the data to be analyzed. The electrochemical measurements is performed at room temperature by implementing the developed readout circuit and 3-electrode electrochemical system which consists of gold sheet with dimensions of 0.25x25x25mm as a working electrode, an Ag/ AgCl as reference electrode (RE) and a platinum wire as counter electrode (CE). The differential pulse voltammetry (DPV) is performed in 40 mL of 10mM of ferricyanide . The experiment setup is shown in the figure 2.

IOP Conf. Series: Materials Science and Engineering **340** (2018) 012016 doi:10.1088/1757-899X/340/1/012016

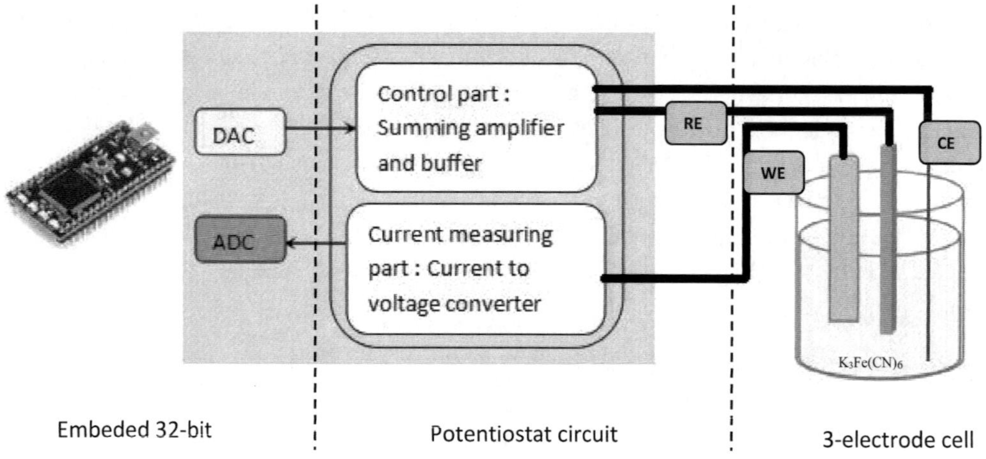

Figure 2 Experiment Setup

The potentiostat will scan the excitation signal between RE and WE and provide the electrochemical reactions to produce a small current through WE and CE. The scan range is set from +0.4V to 1.2V with the scan rate of 0.01V/s. As the scanning reaches the reduction potential, the oxidization rate increased sharply, that raised the sensor current (Faradiac current) and causes a peak in the electrode current. Potentiostat measured this induced current, and after conversion to the voltage, it is sampled through the embedded 12-bit analog-to-digital converters (ADCs) and stored in the MCU memory to be analyzed.

2.2 Design of DPV Algorithms

DPV is a superposition of a staircase and pulse waveform, which the potential between the pulse is not returning back to the same rest potential. Figure 2 shows the flowchart on how the DPV is obtained. Firstly, the program initialize all the parameter values as required. DAC of the microcontroller will produce the excitation voltage and ADC will read the current reading. The parameter values of DPV excitation voltage are as shown in table 1. Theoretically , the pulse period must be set at least twice of the pulse width. The scan rate of DPV is determined by step potential/pulse period.

Table 1 DPV parameter values

Differential Pulse Voltammetry Parameters	Value
Initial Potential(V)	0.4
Final Potential(V)	1.2
Step Potential(V)	0.005
Pulse Width(s)	0.25
Pulse Period(s)	0.50
Pulse Amplitude(V)	0.1
Scan Rate (V/s)	0.01

After the initial potential of 0.4V, the voltage is maintained for 0.25s before it increase to the pulse potential with the increment of 0.1V. Before the potential pulse is applied, the current reading is taken known as base current, Ibase. The pulse potential is increased and maintained for 0.25s and before the pulse ends, the pulse current, Ipulse is sampled. The difference of the two current responses (Ipulse-Ibase) will give a point to produce a peaked voltammogram with the expected peak occured at 0.8V. The new cycle will add the potential from the previous base potential and again repeat the steps. The looping will stop once it reaches the final base potential of 1.2V.

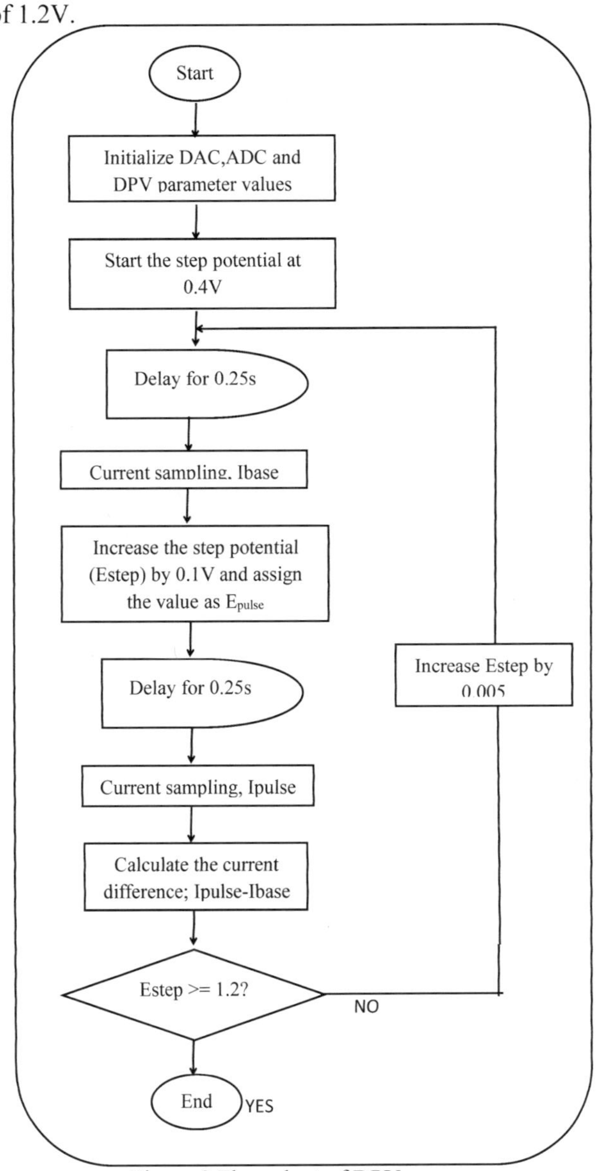

Figure 3 Flow chart of DPV

3. Result and Discussion

To test the performance of the device and its software algorithm, the chemical experiments was conducted on 10mM of Ferricyanide. The result of the voltammogram produced from the electrochemical reaction will be compared with the available commercial potentiostat as a reference. Potential range measurement of the designed instrument is between 0.4V and 1.2V. Figure 3 shows the output from the DAC of the ARM Microcontroller which showing the DPV voltammetry technique as programmed based on the written algorithm. The programmed and actual values of the excitation voltage paramaters is compared and the error percentage is shown in table 2. From the recorded data, the percentage error is very small which is only 1.4%. In terms of timing, the 32-bit embed gives a good response and gives less than 1% of error in completing the scanning time.

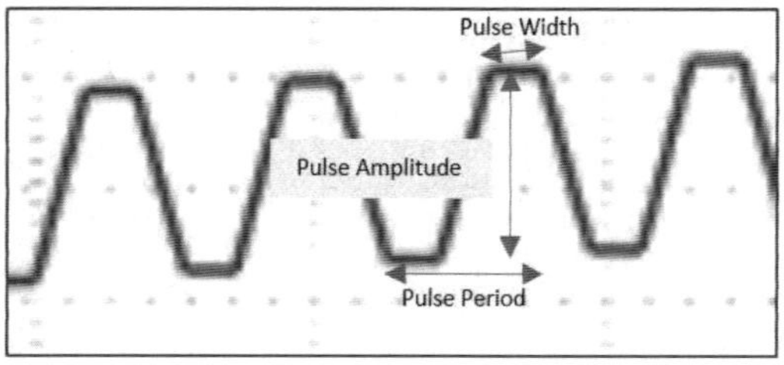

Figure 4 Output of ARM Microcontroller showing DPV voltammetry technique as programmed based on the written algorithm

Table 2 Performance of the generated DPV signal

Parameters	Program	Actual	Percentage Error(%)
Initial Potential(V)	0.400	0.400	0
Final Potential(V)	1.210	1.210	0
Step Potential(V)	0.005	0.0051	1.4
Pulse Width(s)	0.25	0.25	0
Pulse Period(s)	0.50	0.50	0
Pulse Amplitude(V)	0.1	0.0991	0.88
Scan Rate (V/s)	0.01	0.01	0
Completion Time of Scanning(s)	80	80.63	0.78

To validate the developed system, the 3-electrode sensor is used with an authorized potentiostat, Autolab PGSTAT204 to perform electrochemical detection test on 10mM ferricyanide . The DPV parameters value is maintained as in the developed system. Figure 5 shows the graph plotted from the data collected in spreadsheet. As can be seen in the graph, the oxidation peak occur at the voltage of 0.8V when using the commercial potentiostat while the developed device produce the peak in the range of 0.75V and 0.8V; which is 6.25% of reading difference. If required, the design could be calibrated to produce the same result as the commercial potentiostat.

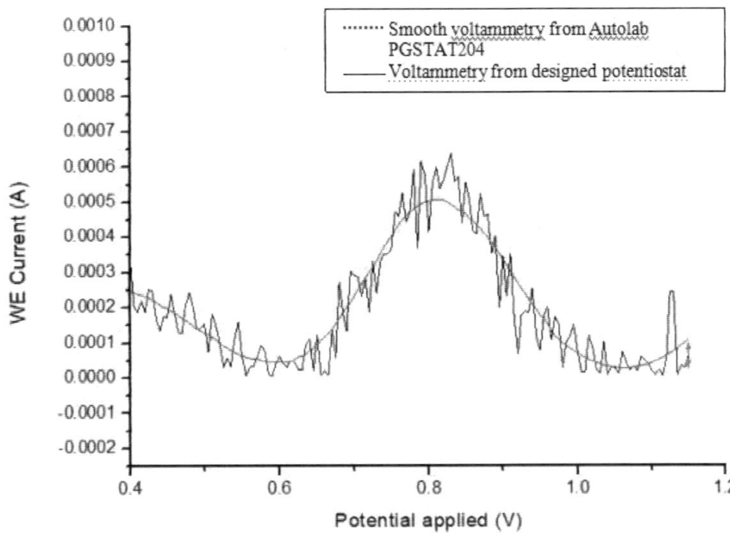

Figure5 voltammogram of 10mM ferricyanide from the designed potentiostat give the same shape with small difference in magnitude compared to the smooth curve output from the commercial potentiostat.

4. Conclusion

The development of 32-bit embedded algorithm using an ARM cortex processor with new developed potentiostat design for in-situ 3-electrode cell sensing were presented in this paper. By demonstrating that the device can produce the voltammogram of 10mM ferricyanide as expected, the capability of the developed device and its software algorithms to detect the redox reaction occur in chemical samples is possible. The device can generate an excitation signal of DPV from 0.4V to 1.2V and produced a peaked voltammogram with relatively small error compared to the commercial potentiostat. The design of potentiostat device and its DPV algorithm is verified.

Acknowledgements

The authors would like to thank Universiti Teknologi MARA, UiTM for all the research facilities. The work is partially supported by Ministry of Education Malaysia under the Niche Research Grant scheme (Project Code: 600- RMI/NRGS 5/3(6/2013) and Collaborative Research in Engineering, Science and Technology (CREST) (Project Code :100-RMI/PRI 16/6/2 (5/2015).

References

[1] M. D. Steinberg, P. Kassal, I. Kerekovi, and I. Murkovi, "Talanta A wireless potentiostat for mobile chemical sensing and biosensing," vol. 143, pp. 178–183, 2015.
[2] A. F. D. Cruz, N. Norena, A. Kaushik, and S. Bhansali, "A low-cost miniaturized potentiostat for point-of-care diagnosis," *Biosens. Bioelectron.*, vol. 62, pp. 249–254, 2014.
[3] C. Huang, W. Li, and H. Lin, "Design and Implementation of a Standalone Potentiostat for Electrochemical Biosensors," vol. 3, no. 2, pp. 5–9, 2013.
[4] Amin, M., Isa, M. M., & Mohd, R. (2015). An Embedded Processing of Differential Pulse Voltammetry (DPV) Data Using ARM processor (LPC1768), 80–84.
[5] M. Azimzadeh, M. Rahaie, N. Nasirizadeh, K. Ashtari, and H. Naderi-Manesh, "An electrochemical nanobiosensor for plasma miRNA-155, based on graphene oxide and gold nanorod, for early detection of breast cancer," *Biosens. Bioelectron.*, vol. 77, pp. 99–106, 2016.
[6] N. A. Mansor, Z. M. Zain, H. H. Hamzah, M. S. A. Noorden, S. S. Jaapar, V. Beni, and Z. H. Ibupoto, "Detection of Breast Cancer 1 (BRCA1) Gene Using an Electrochemical DNA Biosensor Based on Immobilized ZnO Nanowires," *Open J. Appl. Biosens.*, vol. 3, no. 2, pp. 9–17, 2014.
[7] A. J. Veloso, R. R. Fulthorpe, and K. Kerman, "Electrochemistry Communications Electrochemical detection of perchloroethylene using differential pulse voltammetry," vol. 37, pp. 28–31, 2013.
[8] J. Wang, Analytical Electrochemistry, 2nd Edition, vol. I, no. 2. 2001

Integrated Circuit Design of 3 Electrode Sensing System Using Two-Stage Operational Amplifier

S.Rani[1] , W.F.H. Abdullah[1], Z.M.Zain[2], Aqmar N.Z.N[1]

[1] Faculty of Electrical Engineering, Universiti Teknologi MARA, Shah Alam Selangor
[2] Faculty of Applied Science, Universiti Teknologi MARA, Shah Alam Selangor

E-mail: syafiqahrani1990@gmail.com

Abstract This paper presents the design of a two-stage operational amplifier(op amp) for 3-electrode sensing system readout circuits. The designs have been simulated using 0.13µm CMOS technology from Silterra (Malaysia) with Mentor graphics tools. The purpose of this projects is mainly to design a miniature interfacing circuit to detect the redox reaction in the form of current using standard analog modules. The potentiostat consists of several op amps combined together in order to analyse the signal coming from the 3-electrode sensing system. This op amp design will be used in potentiostat circuit device and to analyse the functionality for each module of the system.

Keywords -) two-stage operational amplifier, 3-electrode sensing system, readout circuit, potentiostat

1. Introduction

Electrochemical sensors have encountered rapid growth over the past decades. They are being increasingly and widely researched due to their functionality and ability to detect a variety of target substance or chemical [1]. Moreover, the demand of these sensors are increasing in everyday appliances due to its compact and portable features [1], [2].

The working principle of an electrochemical sensor is to measure and analyse certain variables of an electrochemical cell. This is purely done by analysing the difference in potential, current, or concentration by using 3-electrode sensing system. Reference electrode, counter/auxiliary electrode, and the working electrode are the key components that made up the 3-electrode system. Each electrode has its own unique purpose to the system. Reference electrode (RE) points to the electrode that has an established electrode potential. Meanwhile, counter electrode (CE) is responsible in ensuring that the current will not be able to get through the reference electrode. Along with the working electrode (WE), which is the medium of transportation for the electrons, the 3-electrode system will eventually produce the redox reaction inside the electrochemical cell [1], [3]. The redox reaction occurs when there is an electron transfer from one space to another. The 3-electrode system work together as the potential of a WE is sustained at the same level compared to the RE by adjusting and balancing the current at CE [4]. When the potential is applied to the system, the electrons will be transferred from one electrode to another inside the solution, which in turn will produce the current through the system.

In order to detect and thoroughly analyse the redox reaction from the electrochemical sensor, readout circuit is needed. Thus, the aim for this whole project is to design a potentiostat which will detect the redox reaction in the form of current using standard analog modules. This paper will focus on the design of op amp whereby it consists of several op amps combined together in order to analyse the signal coming from the 3-electrode sensing system. Figure 1 shows basic 3-electrode sensing system. Differential pulse voltammetry (DPV) is an analytical technique that will be used to analyse the redox signal from the input of 3-electrode system.

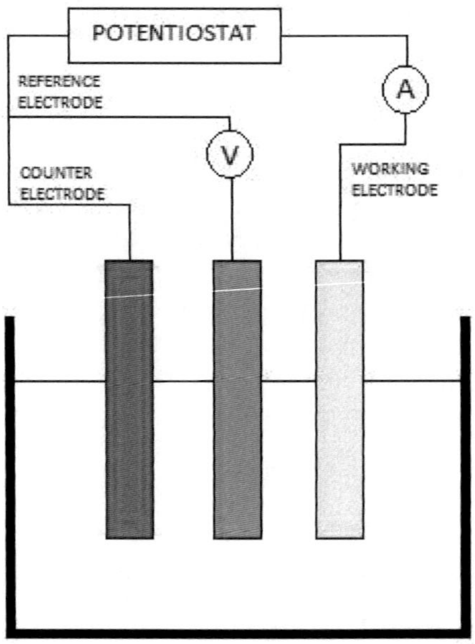

Figure 1 : Basic 3-electrode sensing system.

2. Design Methodology

In the study of modern electrochemical, potentiostat plays the pivotal role in it. Potentiostat is an interfacing circuit that will allow the varying of input voltage and measurement of current. Figure 2 shows the potentiostat design circuit that consists of several op amps combined together. This circuit is divided into several stages. The first stage involves the signal generator from microcontroller through its Digital to Analog (DAC). It will act as a voltage sweeping generator in order for this circuit to be able to receive DPV from microcontroller and send the current feedback of electrochemical reaction to microcontroller. The second stages is to control and maintain the stability of electrochemical reaction of the system. There are using two types of op amp configuration which are summing amplifier and voltage follower. Counter electrode (CE) is connected to the output of the summing amplifier. Main function of this op amp is to ensure the CE supplies required potential and provide a feedback signal in the potentiostat circuit. Meanwhile, the function of voltage follower is to amplify signal coming from the reference electrode (RE) in order to maintain the RE signal current and to ensure there is no current transfer occur at RE. Third and final stage consists of current to voltage converter [5] [6]. At this stage, the current signal from the working electrode (WE) will be converted to voltage signal using current to voltage converter so that it can be read by microcontroller.

IOP Conf. Series: Materials Science and Engineering **340** (2018) 012017 doi:10.1088/1757-899X/340/1/012017

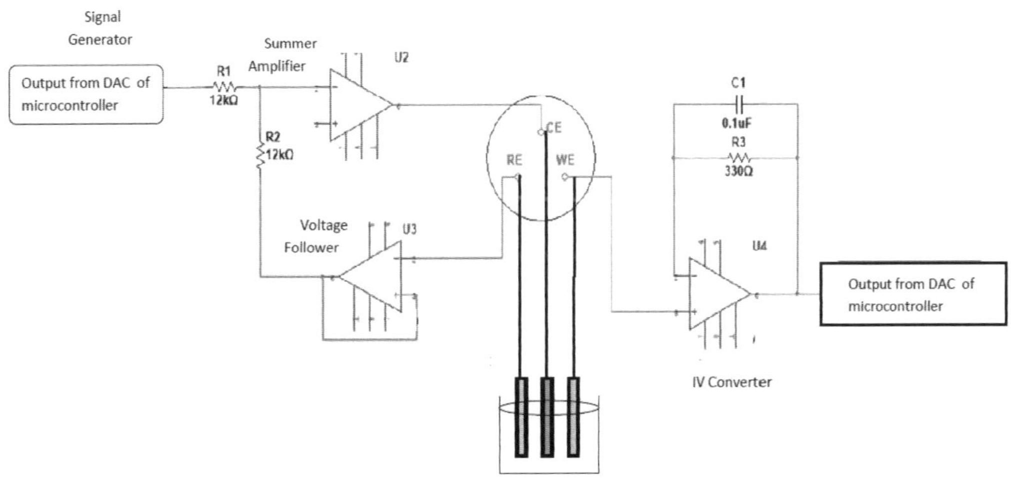

Figure 2 : The design of potentiostat circuit

This design of potentiostat will be implemented in the Intergrated Circuit (IC) design to achieve the aim of this project which is the miniaturisation of potentiostat. In order to do so, a specific op amp needs to be designed at the first place. This op amp will be designed to follow the functionality of the potentiostat such as detection of low current. The two-stage op amp architecture is selected as it complies with the specification needed. Figure 3 below shows the schematic of the two-stage op amp design. The designs have been designed and simulated using 0.13μm CMOS technology from Silterra (Malaysia) with Mentor graphics tools.

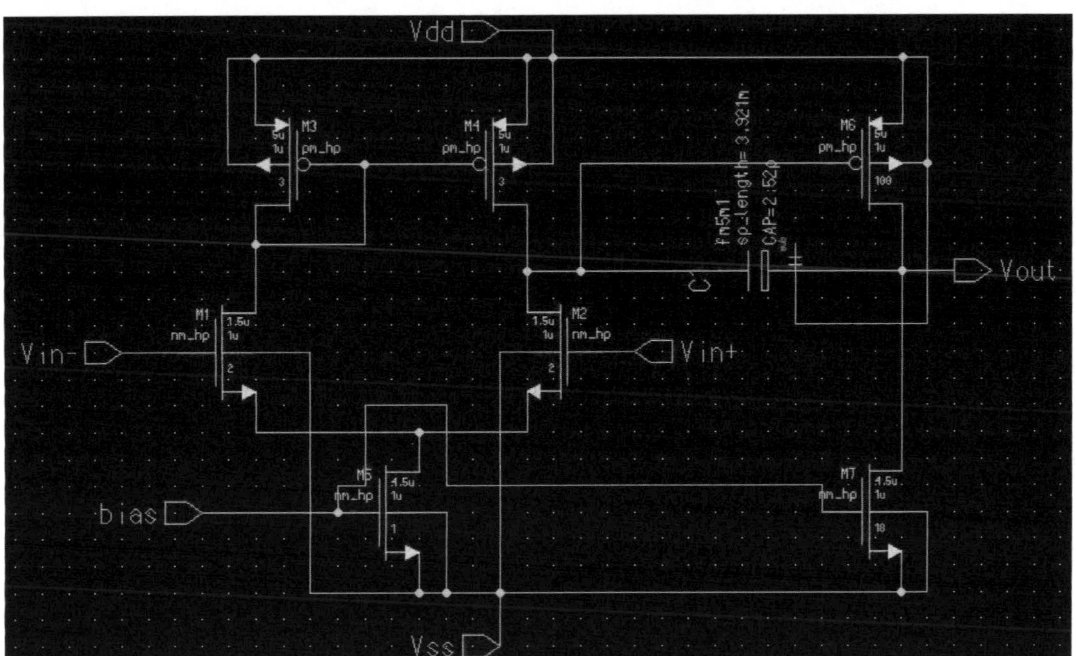

Figure 3 : Schematics of two stage op amp design

3. Result and Discussion

This section discusses the result simulation of two stage op amp design. Table 1 below shows the proposed op amp specification that is suitable for the potentiostat design.

Table 1 : The proposed design of op amp

No	Specification	Proposed Value
1.	Gain	≥70dB
2.	Gain Bandwidth (GB)	≥5MHz
3.	Phase Margin (PM)	≥45°
4.	Slew Rate (SR)	≥5V/μs
6.	Power Dissipation	≤≤2mW

The simulation results for the two-stage amplifier gain are shown in Figure 4. From the figure, the circuit is able to archive a maximum gain of 85.04 dB with a frequency bandwidth of 11.6 MHz. The phase margin which is also in the figure below is 69.135° whereby it meets the requirement of design specification. The slew rate is 5.2V/μs and the power consumption is 199.57μWatt which also comply with the proposed specification. Both outcome for the slew rate and power consumption can be seen in Figure 5 and Figure 6.

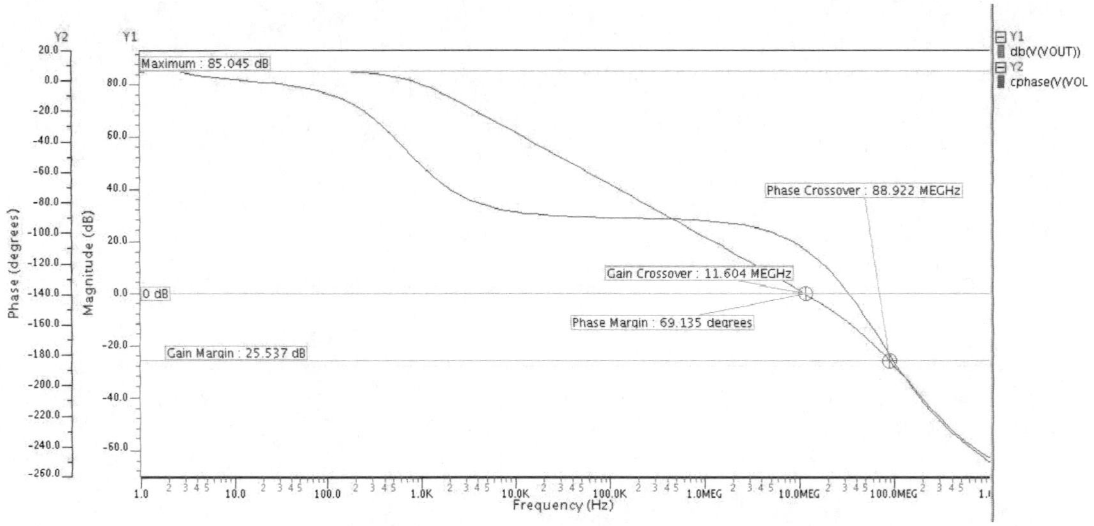

Figure 4: Frequency result of two stage op amp

IOP Conf. Series: Materials Science and Engineering **340** (2018) 012017 doi:10.1088/1757-899X/340/1/012017

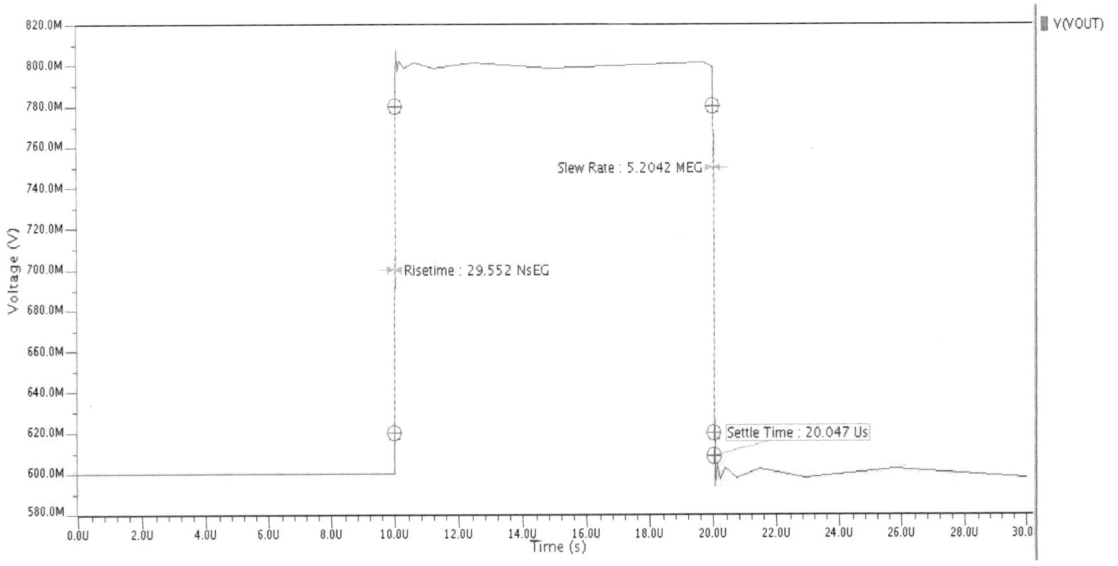

Figure 5: Transient analysis simulation result

Figure 6: Total power dissipation from simulation result

The simulation results in the Table 2 shows that the design meet the requirement as per predetermined in the Table 1.

Table 2 : The comparison proposed and simulation value.

No	Specification	Proposed Value	Simulation value
1.	Gain	≥70dB	85.04 dB
2.	Gain Bandwidth (GB)	≥5MHz	11.6 MHz
3.	Phase Margin (PM)	≥45°	69.135°
4.	Slew Rate (SR)	≥5V/µs	5.2V/µs
6.	Power Dissipation	≤≤2mW	0.19957m Watt

IOP Conf. Series: Materials Science and Engineering **340** (2018) 012017 doi:10.1088/1757-899X/340/1/012017

Figure 7 below shows the layout design for the two stage op amp of using 0.13µm CMOS technology from Silterra (Malaysia) with Mentor graphics tools.

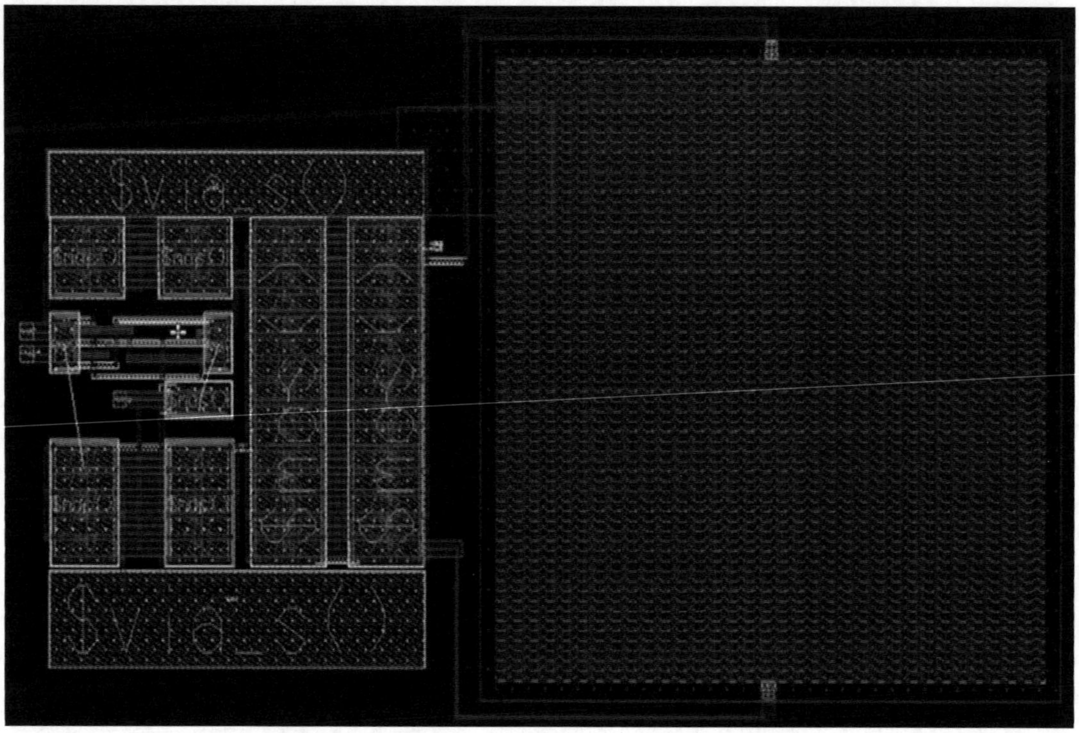

Figure 7 : Layout design for two stage op amp

4. Conclusion

In this paper, the design of a two-stage op amp for 3-electrode sensing system readout circuits was tested through simulations and the results were presented in terms of figure and table. The results, were validated by comparing with the predetermined specifications which are suitable for the potentiostat circuit design.

Acknowledgements

The authors would like to thank Universiti Teknologi MARA, UiTM for all the research facilities. The work is partially supported by Ministry of Education Malaysia under the Niche Research Grant scheme (Project Code: 600- RMI/NRGS 5/3(6/2013) and Collaborative Research in Engineering, Science and Technology (CREST) (Project Code :100-RMI/PRI 16/6/2 (5/2015).

IOP Conf. Series: Materials Science and Engineering **340** (2018) 012017 doi:10.1088/1757-899X/340/1/012017

References

[1] Y. Cui, "Electronic Materials, Devices, and Signals in Electrochemical Sensors," in IEEE Transactions on Electron Devices, vol. 64, no. 6, pp. 2467-2477, 2017.

[2] L. Lombardo, S. Generelli, N. Tscharner, D. Migliorelli and N. Donato, "A compact electronic interface for electrochemical sensors," 2016 IEEE Sensors Applications Symposium (SAS), Catania, 2016, pp. 1-4.

[3] Devengenzo, M. (n.d.). Standard Electrodes. Retrieved August 7, 2015, https://chem.libretexts.org/Core/Analytical_Chemistry/Electrochemistry/Electrodes/Standard_Hydrogen_Electrode

[4] Karlheinz Kellner, Thomas Posnicek, Jörg Ettenauer, Karen Zuser, Martin Brandl, A New, Low-cost Potentiostat for Environmental Measurements with an Easy-to-use PC Interface, Procedia Engineering, Volume 120, 2015, Pages 956-960, ISSN 1877-7058.

[5] Loncaric, C. (2011). Design, Construction, And Testing Of A Compact Usb Powered Potentiostat For Biosensor Applications by Carlyn Loncaric Thesis Submitted In Partial Fulfillment Of The Requirements For The Degree Of Master Of Applied Science In the School of Engineering Sc.

[6] Oluwole, O. O., Adegoke, T. O., & Ajide, O. O. (2014). Development Of A Field-Portable Digital Potentiostat, 5(4), 654

IOP Conf. Series: Materials Science and Engineering **340** (2018) 012018 doi:10.1088/1757-899X/340/1/012018

Drying Temperature Dependence of Sol-gel Spin Coated Bilayer Composite ZnO/TiO$_2$ Thin Films for Extended Gate Field Effect Transistor pH Sensor

R A Rahman[1], M A Zulkefle[1], K A Yusoff[1], W F H Abdullah[2], M Rusop[1,2,3], S H Herman[*1,2,4]

[1]NANO-Technology Centre (NET), Faculty of Electrical Engineering, Universiti Teknologi MARA (UiTM), 40450 Shah Alam, Selangor, Malaysia
[2]Faculty of Electrical Engineering, Universiti Teknologi MARA (UiTM), 40450 Shah Alam, Selangor, Malaysia
[3]NANO-Science Center (NST), Institute of Science (IOS), Faculty of Applied Sciences, Universiti Teknologi MARA (UiTM), 40450 Shah Alam, Selangor, Malaysia
[4]Microwave Research Institute, University Teknologi MARA, 40450 Shah Alam, Selangor, Malaysia

*Corresponding Author: hana1617@salam.uitm.edu.my

Abstract. This study presents an investigation on zinc oxide (ZnO) and titanium dioxide (TiO$_2$) bilayer film applied as the sensing membrane for extended-gate field effect transistor (EGFET) for pH sensing application. The influences of the drying temperatures on the pH sensing capability of ZnO/TiO$_2$ were investigated. The sensing performance of the thin films were measured by connecting the thin film to a commercial MOSFET to form the extended gates. By varying the drying temperature, we found that the ZnO/TiO$_2$ thin film dried at 150°C gave the highest sensitivity compared to other drying conditions, with the sensitivity value of 48.80 mV/pH.

1. Introduction

Extended Gate Field Effect Transistor (EGFET) is gaining much interest as an alternative to Ion Selective Field-Effect Transistor (ISFET) [1-3], to isolate the FET from the chemical environment [4]. The extended gate is a chemically-sensitive membrane, deposited on the signal line extended from the FET gate electrode [5]. EGFET have been applied in many applications, especially as pH sensor [6-8]. While ISFET and EGFET have the same surface ion adsorption mechanism for the pH sensing [9], compared with the ISFET, EGFET has many advantages, such as temperature and light insensitivity, low-cost, simpler to passivate and package, and better long-term stability [10].

Metal oxides are being used as the sensing layer for both ISFET and EGFET. The first pH-sensitive membrane used was silicon dioxide (SiO$_2$), which was applied to ISFET [11]. Metal oxides such as others are tin oxide (SnO$_2$) [14][12], ruthenium oxide (RuO$_2$) [5], zinc oxide vanadium oxide (V$_2$O$_5$) [17][13] and carbon-nanotube (CNT) [18][14] were chosen in earlier investigations of EGFET pH sensor. Other than that, titanium dioxide [15-16] and zinc oxide [17-18] also has drawn an attraction to be applied as the sensing materials for EGFET pH sensor. All of these materials were chosen due to their pH-sensitive properties. These sensing materials can be deposited by using various methods such as RF Sputtering [19], thermal chemical vapour deposition (TCVD) [15], RF Magnetron Sputtering [20], sol-gel spin-coating technique [21-22] However, there are numerous study on single metal oxide deposited as the sensing membrane for pH sensor application [22-26]. E. M. Guerra et.al in her study reported that the single TiO_2 used as the sensing membrane for EGFET based pH sensor application produced the sensing capability of 47.8 mV/pH [27]. This sensing membrane was fabricated by using sol-gel spin coating technique.

Additionally, nanostructured ZnO that was prepared by using electroless metal deposition (EMD) and atomic layer deposition (ALD) was also applied as the sensing membrane for EGFET, where this sensing membrane produced 46.25 mV/pH sensitivity value [28]. In the literatures however, there are no reported works regarding usage of ZnO/TiO_2 bilayer structure as a pH sensing membrane. To further increase the sensor sensitivity, in this work, the combination of two metal oxides for the sensing layer, namely ZnO and TiO_2 was explored. Besides the sensing characteristics, the physical properties of the ZnO/TiO_2 bilayer film was also investigated.

2. Methodology

2.1. ZnO and TiO_2 sol-gel preparation

There are few important steps involved in this experimental. The first step was preparation of ZnO solution. In order to obtained 0.4M solution concentration, zinc acetate dehydrate ($Zn(CH_3COO)_22H_2O$) as starting material was dissolved in 2-methoxyethanol solvent and methanolamine (MEA, $C_2H_7N_{14}$) stabilizer. The molar ratio of MEA to zinc acetate was 1.0M. This solution was stirred at 300 rpm while heated 80 °C for 3 hours. TiO_2 sol-gel solution was prepared using absolute ethanol, glacial acetic acid, titanium isopropoxide, Triton X-100 and deionized water to form a 0.1M solution. The detailed recipe for TiO_2 sol-gel solution has been reported elsewhere [29].

2.2. Deposition process

For deposition of the sensing layer, ten drops of TiO_2 solution were deposited on the ITO coated glass substrate followed by ten drops of ZnO solution. In between the first and second layer deposition, the thin film was dried at room temperature for 10 minutes. In this work, the drying temperature was varied from room temperature (as deposited) until 150°C. The deposited bilayer thin films were then annealed for 15 minutes at 400°C. The configuration of the deposited ZnO/TiO_2 bilayer film is shown in Fig. 1 (a) below. The annealed thin film was then connected to a copper wire to form the extended gate of a commercialized FET incorporated in a readout interfacing circuit (ROIC). For pH sensing measurement, the extended gate was dipped in pH buffer solution, with the setup shown in Fig. 1 (b). A commercialized Ag/AgCl reference electrode was employed in the measurement. The thin films were also characterized by atomic force microscope (AFM), field emission scanning electron microscope (FESEM) and also energy dispersive x-ray spectroscopy (EDX).

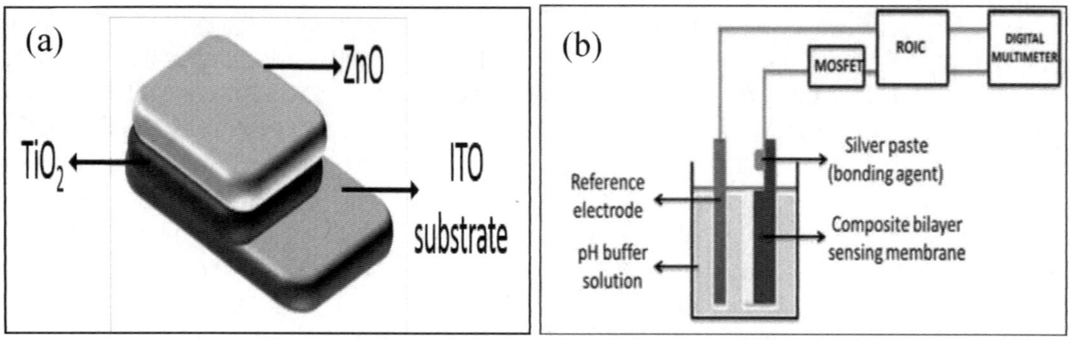

Figure 1 : (a) Zn/TiO_2 composite bilayer thin film configuration deposited by using sol-gel spin coating technique and (b) EGFET pH sensor measurement setup

3. Results and discussion

Fig. 2 shows the pH sensitivity dependence on the drying temperature. The sensitivity is derived from the slope of the ROIC output voltage versus pH graph. It can be seen that the sensitivity increases with the drying temperature. The detailed values are tabulated in Table 1. Table 1 also shows the thickness of each thin film which decreases as the drying temperature increase. The sensitivity values ranged from 35 to 48 mV/pH while the thickness ranged from 28 to 63 nm. The linearity also is higher for thinner film compared to the thicker ones with the highest is 0.985.

The drying or pre-heating process alters the surface of the thin film [2] that influences the roughness and thickness value [30]. Drying process is important to remove the excessive solvent, moisture and residual chemicals contained in the ZnO and TiO_2 solution during the deposition process. In this study, the solvent used are 2-methoxyethanol and absolute ethanol, which have the boiling of 125°C and 78.3°C respectively. It is crucial to find a suitable and optimum drying temperature because the solvent and stabilizer required to be vaporized. In this study, ZnO/TiO_2 bilayer film dried at 150°C produced highest sensitivity due to the completion of the solvent removal during the drying process. If the temperature is lower than these boiling points, the vaporization and removal process are not complete and will be retarded [31]. This condition will lead to the abrupt vaporization during the annealing process [32]. However, if the temperature is too high, it will cause the solvent to vapor abruptly during the drying process.

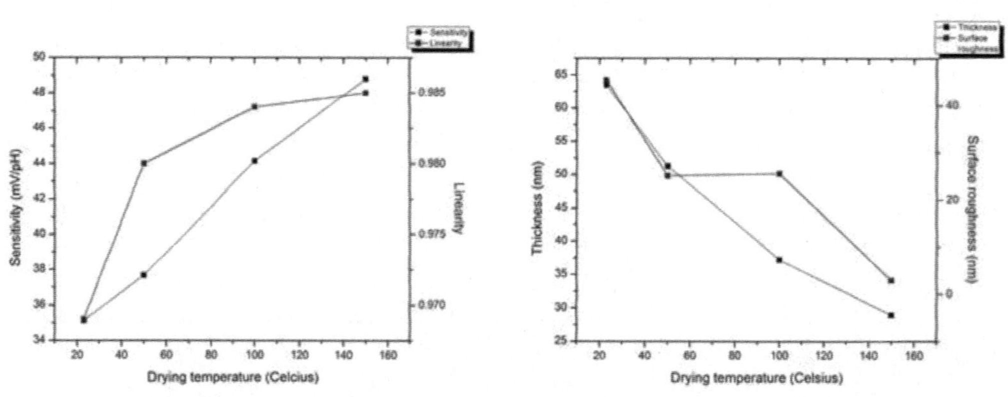

IOP Conf. Series: Materials Science and Engineering **340** (2018) 012018 doi:10.1088/1757-899X/340/1/012018

Figure 2 : (a) Zn/TiO_2 composite bilayer thin film configuration deposited by using sol-gel spin coating technique and (b) EGFET pH sensor measurement setup

Table 1. Sensitivity and linearity values for each bilayer film annealed at different duration

Annealing temperature (°C)	**Sensitivity (mV/pH)**	**Linearity**	**Thickness (nm)**	**Surface roughness (nm)**
Room temperature	35.13	0.9696	63.42	45.36
50	37.67	0.9800	51.29	25.04
100	44.15	0.9840	37.20	25.53
150	48.80	0.9850	28.98	2.90

To prevent this condition, the drying process need to be done under a suitable temperature, and this deposition process will be completed with the annealing process, after the drying process done. It is generally known that annealing improves the quality of the thin films. From Table 1, higher drying temperature resulted in lower surface roughness. From these results it can be said that better quality of the film improves the pH sensing characteristics of the ZnO/TiO_2 bilayer thin films. Fig.3 (a), (b), (c) and (d) are the three dimensional (3D) topographical images of each ZnO/TiO_2 bilayer films obtained from AFM imaging. ZnO/TiO_2 bilayer film dried at highest temperature produced smallest roughness value which is 2.90 nm. In comparison, the roughness value for the bilayer film decrease when the drying temperature was raised up to 150 °C and it can be seen that the thin film is more uniform compared to the others. This thin film exhibited the highest sensitivity and linea rity when it was used as the sensing layer for the EGFET pH sensor.

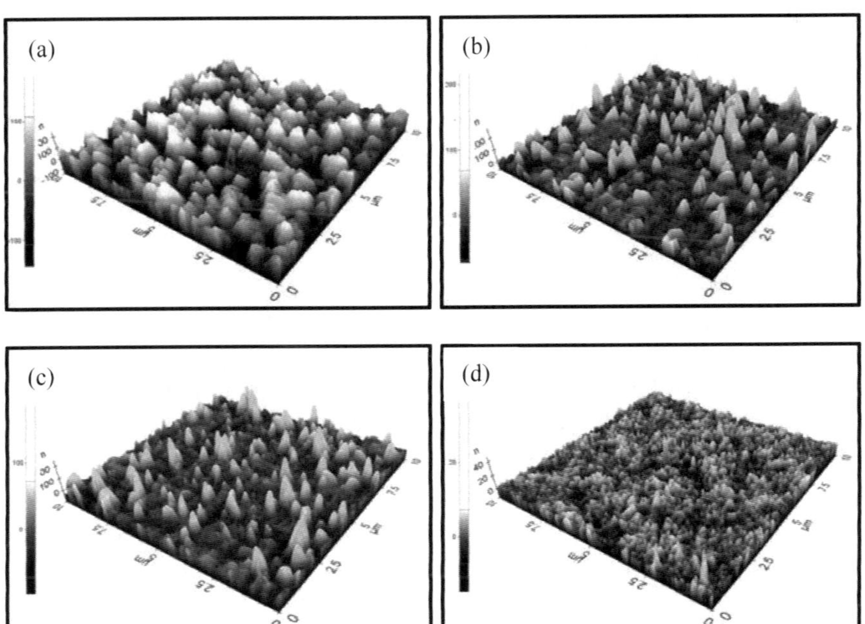

Figure 3 : AFM topography of ZnO/TiO_2 thin film dried at (a) room temperature, (b) 50 °C, (c) 100 °C, and (d) 150 °C respectively.

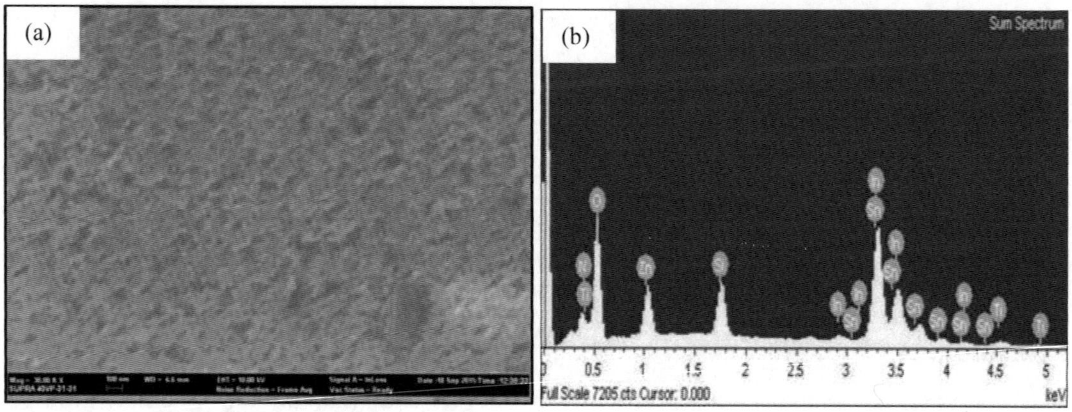

Figure 4 : (a) Surface morphology of ZnO/TiO_2 thin film, observed by FESEM and (b) Elements contained on the thin film, characterized by EDX characterization.

On the other hand, Fig.4 (a) present the surface morphology of the ZnO/TiO_2 bilayer film deposited at 3000 rpm. From the image, it can be seen that ZnO had been deposited successfully with uniform surface structure. In addition, Fig. 4 clearly showed that the entire surface was covered with the ZnO, as the top material for this bilayer film. However, only ZnO surface morphology can be observed by FESEM due to its function as the upper layer for this bilayer film. Other than that, it was confirmed that ZnO and TiO_2 were successfully deposited based on the EDX characterization. The EDX characterization, which shows in Fig. 4 (b) prove that all of the desired materials, which is ZnO and TiO_2 were deposited and exist on the substrate used. EDX result describe that ZnO was dominant, compare with the TiO_2. However, TiO_2 is still dominant with the other elements existed, such as silicon (Si), tin (Sn), indium (In), nitrogen (N), and also oxygen (O_2). Elements such as Si, Sn and In were from the substrate used, ITO coated glass, but elements like N and O_2 might originate from the air ambient and environment contamination during the annealing process.

4. Conclusion

ZnO/TiO_2 bilayer composite thin films have been successfully deposited on ITO coated glass substrate. The deposited bilayer composite thin film has been proven to be sensitive towards pH. The pH sensitivity of the thin films depends on the drying temperature in which, increasing drying temperature improved the sensitivity. The highest sensitivity achieved in this work was 48.80 mV/pH with the linearity of 0.9850. Higher drying temperature process results in better quality of the thin film. Instead of sensing performance, physical properties of ZnO/TiO_2 bilayer film was also improved with the increasing drying temperature, which support the finding from the EGFET measurement.

Acknowledgement
The authors would like to thank all members of NANO-ElecTronic Center (NET) and NANO-Science Center (NST) Universiti Teknologi MARA, UiTM for all the research facilities. The work is partially supported by the Ministry of Higher Education Malaysia under Niche Research Grant Scheme, (600-RMI/NRGS 5/3 (7/2013)).

References
[1] P. C. Yao, J. Chiang, and M. Lee, "Application of sol-gel TiO_2 film for an extended-gate H^+ ion-sensitive field-effect transistor," *Solid State Sciences,* vol. 28, 47-54, 2014.
[2] M. A. Rosdan, N. S. Kamarozaman, W. F. H. Abdullah, M. I. S Yono and S. H. Herman, "Sputtered titanium dioxide thin film for extended-gate FET sensor application," *2013 IEEE International Regional Symposium Micro and Nanoelectronic (RSM),* pp. 219–222, 2013.
[3] L.-L Chi, J.-C Chou, W.-Y Chung, T.-P Sun, and S.-K Hsiung, "Study on extended gate field effect transistor with tin oxide sensing membrane," *Mater. Chem. Phys.* vol. 63, no. 1, pp. 19–23, 2000.
[4] P. Yao, and M. Lee, "Annealing effect of sol-gel TiO_2 thin film on pH-EGFET sensor," *Mater. Chem. Phy*, no. 1, pp. 3–6, 2014.
[5] J. V. D Spiegel, I. Lauks, P.Chan, and D. Babic, "The extended gate chemically sensitive field effect transistor as multi-species microprobe," *Sensors and Actuators,* vol. 4, pp. 291–298, 1983.
[6] C. Toumazou, Georgiou, P. P. Bergveld, "40 years of ISFET technology: From neuronal sensing to DNA sequencing," *Electron. Lett.* vol. 47, no. 26, p. S7, 2011.
[7] P. Zubiate, C. R. Zamarreño, I. D. Villar, I. R. Matias, and F. Arregui, "Tunable optical fiber pH sensors based on TE and TM Lossy Mode Resonances (LMRs)," *Sensors Actuators B Chem.* vol. 231, pp. 484–490, 2016.
[8] L. C. Yen, M. T. Tang, F. Y. Chang, T. M. Pan, T. S. Chao, and C. H. Lee, "Improvement in pH sensitivity of low-temperature polycrystalline-silicon thin-film transistor sensors using H_2 sintering. *Sensors (Switzerland),* vol. 14, no. 3, pp. 3825–3832, 2014.
[9] S. Y. Yang, C. W. Chen, and J. C. Chou, "Investigation on the sensitivity of TiO_2:Ru pH sensor by Taguchi design of experiment," *Solid. State. Electron,* vol. 77, pp. 82–86, 2012.
[10] Y. S. Chien, W. L. Tsai, I. C. Lee, J. C. Chou, and H. C. Cheng, "A novel pH sensor of extended-gate field-effect transistors with laser-irradiated carbon-nanotube network," *IEEE Electron Device Lett.*, vol. 33, no. 11, pp. 1622–1624, 2012.
[11] P. Bergveld, "Thirty years of ISFETOLOGY: What happened in the past 30 years and what may happen in the next 30 years," *Sensors Actuators, B Chem.*, vol. 88, no. 1, pp. 1–20, 2003.
[12] M. A. Zulkefle, R. A. Rahman, K. A. Yusof, W. F. H. Abdullah, M. Rusop, and S. H. Herman, "Annealing time dependence of the physical, electrical and pH response characteristics of spin coated TiO_2 thin films," *IOP Conf. Ser. Mater. Sci. Eng.,* vol. 99, p. 012021, 2015.
[13] M . A. Zulkefle, R. A. Rahman, K. A. Yusof, W. F. H. Abdullah, M. Rusop, and S. H. Herman, "Sensing capability of TiO_2 thin films with different thicknesses as sensing membrane of EGFET pH sensor," *2015 International Symposium on Technology Management and Emerging Technologies (ISTMET),* pp. 30–33, 2015.
[14] L. Chi, J. Chou, W. Chung, T. Sun, and S. Hsiung, "Study on extended gate field effect transistor with tin oxide sensing membrane," *Mater. Chem. Phys.,* vol. 63, pp. 19–23, 2000.
[15] A. B. Rosli, N. D. H. A. Patah, M. A. Rosdan, M. M. Marbie, M. H. Juhari, S. S. Shariffudin, S. H. Herman, and M. Rusop, "Influence of metal catalyst for zinc oxide nanostructures grown by TCVD method for extended-gate FET sensor application," *RSM 2013 IEEE Reg. Symp. Micro Nanoelectron,* pp. 171–174, 2013.
[16] A. Fulati, S. M.; Usman Ali, M. Riaz, G. Amin, O. Nur, and M. Willander, "Miniaturized pH sensors based on zinc oxide nanotubes/nanorods," *Sensors*, vol. 9, no. 11, pp. 8911–8923, 2009.
[17] W. Tsai, B. Huang, K. Wang, Y. Huang, P. Yang, and H. Cheng, "Functionalized Carbon Nanotube Thin Films as the pH Sensing Membranes of Extended-Gate Field-Effect Transistors

on the Flexible Substrates." *IEEE Transaction on Nanotechnology.* vol. 13, no. 4, pp. 760–766, 2014.

[18] G. M. Ali, R. H. Dhaher, and A. Abdullateef, "pH sensing characteristics of EGFET based on Pd-doped ZnO thin films synthesized by sol-gel method," *2015 Third Int. Conf. Technol. Adv. Electr. Electron. Comput. Eng*, vol. 4, no. 2, pp. 234–238, 2015

[19] A. Sardarinejad, D. Maurya, and K. Alameh, "The pH Sensing Properties of RF Sputtered RuO_2 Thin-Film Prepared Using Different Ar/O_2 Flow Ratio," *Materials (Basel),* vol. 8, no. 6, pp. 3352–3363, 2015.

[20] D. Maurya, A. Sardarinejad, and K. Alameh, "Recent Developments in R.F. magnetron sputtered thin tilms for pH sensing applications—An Overview," *Coatings,* vol. 4, no. 4, pp. 756–771, 2014.

[21] S. Islam, N. Bidin, S. Riaz, G. Krishnan, and S. Naseem, "Sol–gel based fiber optic pH nanosensor: Structural and sensing properties," *Sensors Actuators A Phys.* vol. 238, pp. 8–18, 2016.

[22] G. Owens, R. K. Singh, F. Foroutan, M. Alqaysi, C.-M. Han, C. Mahapatra, H.-W. Kim, and J. C. Knowles, "Sol-gel Based Materials for Biomedical Applications," *Prog. Mater. Sci.* vol. 77, pp. 1–79, 2016.

[23] J. Li, S. Chang, S. Chang, and T. Tsai, "Sensitivity of EGFET pH Sensors with TiO_2 Nanowires," *ECS Solid State Letters*, vol. 3, no. 10, pp. 123–126, 2014.

[24] C. S. Lee, S. K. Kim, and M. Kim, "Ion-sensitive field-effect transistor for biological sensing," *Sensors*, vol. 9, no. 9, pp. 7111–7131, 2009.

[25] H. José, N. Pedroza, D. Mello, and M. Mulato, "Well-established materials in microelectronic devices systems for differential-mode extended-gate field effect transistor chemical sensors," *Microelectronic Enginieering*, vol. 160, pp. 73–80, 2016.

[26] R. Rahimi, M. Ochoa, T. Parupudi, X. Zhao, I. Yazdi, R. Mehmet, A. Tamayol, A. Khademhosseini, and B. Ziaie, "A low-cost flexible pH sensor array for wound assessment," *Sensors Actuators B. Chem.* vol. 229, pp. 609–617, 2016.

[27] E. M. Guerra, and M. Mulato, "Titanium Oxide Nanorods pH Sensors : Comparison between Voltammetry and Extended Gate Field Effect Transistor Measurements," *Material Sciemce and Applications,* pp. 459–466, 2014.

[28] H. H. Li, C. E. Yang, C. C. Kei, C. Y. Su, W. S. Dai, J. K. Tseng, P. Y. Yang, J. C. Chou, and H. C. Cheng, "Coaxial-structured ZnO/silicon nanowires extended-gate field-effect transistor as pH sensor," *Thin Solid Films*, vol. 529, pp. 173–176, 2013.

[29] R. A. Rahman, M. A. Zulkefle, W. F. H. Abdullah, M. Rusop, and S. H. Herman, "Annealing temperature effect on the electrical and pH Sensitivity of TiO_2/ZnO Bilayer Film," *International Symposium Technology on Management and Emerging Technologies (ISTMET)*, vol. 012019, pp. 39-43, 2015.

[30] I. H. H. Affendi, Z. Nurbaya, N. E. A. Azhar, M. N. Wahida, P. Sarah, and M. Rusop, "Effect of annealing and drying temperature on TiO_2 nanostructured film," *2nd International Conference on Electronic Design (ICED)*, pp. 1–4, 2014.

[31] S. S. Shariffudin, M. H. Mamat, S. H. Herman, and M. Rusop, "Influence of drying temperature on the structural, optical, and electrical properties of layer-by-layer ZnO nanoparticles seeded catalyst," *J. Nanomater.* vol. 2012, 2012.

[32] A. Pakdel, and F. E. Ghodsi, "Influence of drying conditions on the optical and structural properties of sol–gel-derived ZnO nanocrystalline films," *Pramana*, vol. 76, no. 6, pp. 973–983, 2011.

IOP Conf. Series: Materials Science and Engineering **340** (2018) 012019 doi:10.1088/1757-899X/340/1/012019

Annealing Temperature Dependence of ZnO Nanostructures Grown by Facile Chemical Bath Deposition for EGFET pH Sensors

Aimi Bazilah Rosli[1,2], Zaiki Awang[3], Shafinaz Sobihana Shariffudin[1] and Sukreen Hana Herman[1,2,3]

[1]NANO-ElecTronic Centre, Faculty of Electrical Engineering, Universiti Teknologi MARA, 40450 Shah Alam, Selangor, Malaysia.

[2]Integrated Sensors Research Group, Faculty of Electrical Engineering, Universiti Teknologi MARA, 40450 Shah Alam, Selangor, Malaysia

[3]Microwave Research Institute, Universiti Teknologi MARA, 40450 Shah Alam, Selangor, Malaysia

Corresponding author : hana1617@salam.uitm.edu.my

Abstract. Zinc Oxide (ZnO) nanostructures were deposited using chemical bath deposition (CBD) technique in water bath at 95 °C for 4 h. Post-deposition heat treatment in air ambient at various temperature ranging from 200-600 °C for 30 min was applied in order to enhance the electrical properties of ZnO nanostructures as the sensing membrane of extended-gate field effect transistor (EGFET) pH sensor. The as-deposited sample was prepared for comparison. The samples were characterized in terms of physical and sensing properties. FESEM images showed that scattered ZnO nanorods were formed for the as-deposited sample, and the morphology of the ZnO nanorods changed to ZnO nanoflowers when the heat treatment was applied from 200-600 °C. For sensing properties, the samples heated at 300 °C showed the higher sensitivity which was 39.9 mV/pH with the linearity of 0.9792. The sensing properties was increased with the increasing annealing treatment temperature up to 300 °C before decreased drastically.

1. Introduction

pH plays a critical role in regulating the reactivity of many chemical species, thus its measurement is fundamental in many fields such as environmental science, agriculture, chemical engineering, chemistry, food science, biology and medicine. Field Effect Transistor (FET) based devices have attracted a lot interest in pH sensors field. Ion-sensitive field-effect transistor (ISFET) is one of FET biosensors that was introduced by Bergveld in 1970 [1-3]. The ISFET used a standard MOSFET configuration except that the gate was replaced with sensing membrane that was directly exposed to the solution [4-6]. This configuration induced device instability and also low current sensitivity of ISFET [7]. In 1983, extended gate field-effect transistor (EGFET) was proposed by Van der Spiegel et al to overcome the disadvantages of ISFET. EGFET is a structure used to isolate FET from the chemical environment [4, 8] in which, a chemically sensitive membrane is connected at the end of the signal line extended from the FET gate electrode [9]. EGFET has many advantages compared to ISFET, including insensitivity to light, simplicity of passivation and packaging [10-12]. EGFET uses the same principle of operation of the ISFET with the principal distinction between pH-ISFET and pH-EGFET sensors is the impedance of sensing films [11]. Metal oxide semiconductors have attracted considerable attention as the EGFET sensing membranes due to their high sensitivity, fast response and recovery, low detection limits and low fabrication cost as sensors [13-15]. At present, ZnO, SnO_2 and TiO_2 have been found to be prominent sensing materials by scientific community [14, 16, 17]. Among these materials, extensive studies on the ZnO have been done for various applications due to its excellent properties in terms of electrical and optical [18, 19] properties. Besides that, ZnO also exhibits various types of morphology such as nanorods, nanowire, tetrapods and nanocombs [20-23]. These nanostructures were reported to increase the sensitivity of sensors due to its high surface to volume ratio that leads to high surface area for sensor activity [24]. Various methods have been developed to attain various structures of ZnO such as RF sputtering, thermal chemical vapor (TCVD), hydrothermal and sol-gel process [11, 25-27]. Among these methodology, the solution-based process is favorable due to its low cost, flexibility when it comes to tuning the reaction parameters, and easy to handle[28, 29].

In this work, the ZnO nanostructures was deposited using simple chemical bath deposition (CBD) process in water bath at 95 °C for 30 min. The post-deposition heat treatment temperature was varied from 200 to 600 °C to study the effect of annealing temperature on ZnO nanostrucutres as EGFET pH sensors.

2. Experimental details

2.1. *ZnO Nanostructures Deposition*

The ZnO nanostructures were grown by simple the chemical bath deposition process. For the solution preparation, 0.23 g zinc nitrate hexahydrate [$Zn(N0_3)_2 . 6H_2O$] was mixed with 0.11 g of hexamethylenetetramine ($C_6H_{12}N_4$) in 100 ml of deionized water. Then, the mixture was stirred on hot plate at 300 rpm for 3 hours at room temperature to ensure well dispersion. After that, the mixture was transferred into a water bath with the constant temperature 95 °C for 4 hours followed by a 14-hours cooling process at room temperature. Then, the sample was rinsed with deionized (DI) water to remove the unused salt. After the drying process, the samples were annealed in at the temperature ranging from 200 - 600 °C for 30 min. As-deposited sample was also prepared as comparison.

2.2. *Sample Characterization*

The surface morphology of the ZnO nanostructure was examined by field- emission scanning electron microscope (FESEM, JSM- 7600F). For pH sensing capability, the sensitivity measurement was taken using EGFET setup equipment as shown in Fig.1. ZnO nanostructure as the pH sensitive membrane and reference electrode was connected to a commercial metal- oxide semiconductor FET (MOSFET), which is connected to a readout interface circuit (ROIC). The sensitivity of ZnO thin film towards the pH buffer solution was measured by dipping the sensing membrane in pH buffer solution ranging from 4 to 12. The sensitivity values were obtained from the slope of the plotted graph V_G versus pH level. Detailed measurement setup is described elsewhere [30].

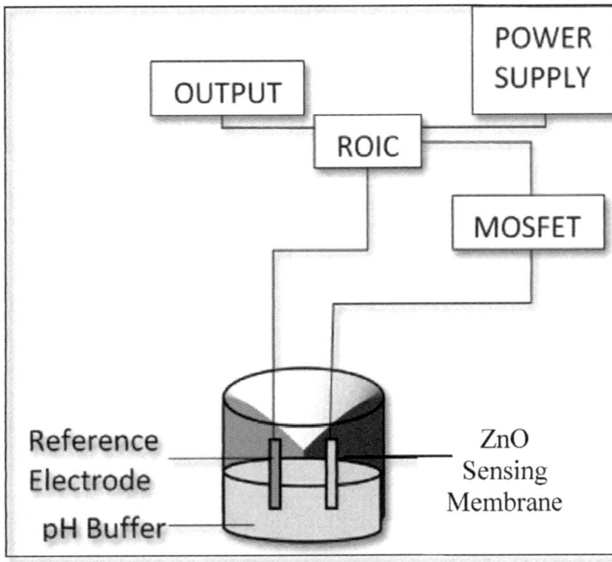

Fig.1 Experiment setup for pH sensitivity measurement.

3. Results and discussions

3.1. ZnO Growth Morphology

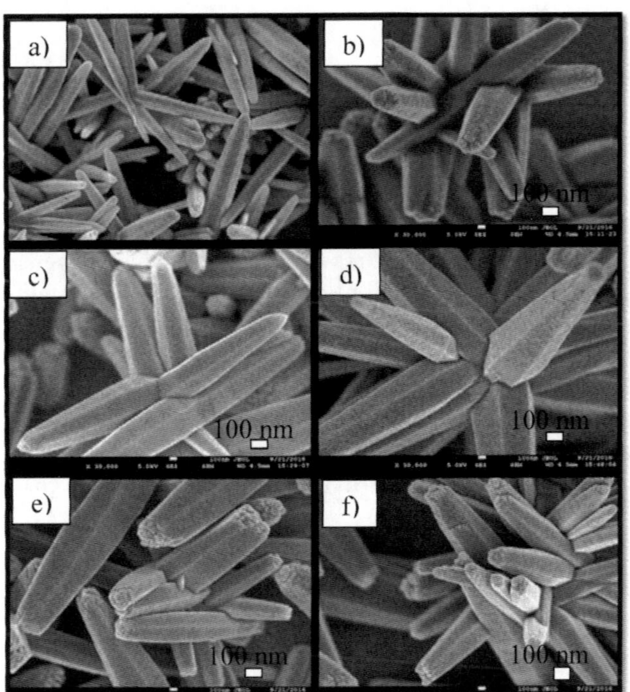

Fig.2 FESEM images of ZnO nanostructures deposited at b) 200 c) 300 d) 400 e) 500 and f) 600 °C post annealing treatment. a) shows the as-deposited sample.

The surface morphology of ZnO nanstructures for as-deposited sample and annealed samples at various annealing temperatures are shown in Fig.2. From the figure, it can be seen that the as-deposited sample (Fig.2 (a) is ZnO nanorods. After the post-deposition annealing treatment at 200 to 600 °C, the ZnO nanorods agglomerated and formed the ZnO nanoflowers. Increasing the post annealing temperature caused the ZnO nanoflowers diameter and length to increase. Annealing process are known to improve the crystallinity of a material and at the same time may influence the growth direction. The increase in the annealing temperature means that higher energy received by the atoms to rearrange themselves [31] thus resulting in longer rod-petals.

3.2. pH Sensing Behaviour

Table 1. The value of sensitivity and linearity for each sample annealed at various temperature

	Sensitivity (mV/pH)	Linearity
As-deposited	3.4	0.0683
200 °C	29.2	0.8476
300 °C	39.8	0.9672
400 °C	25.0	0.9658
500 °C	12.1	0.652
600 °C	19.0	0.8434

The sensitivity and linearity of the as deposited sample and annealed samples at various annealing time are shown in Table 1. The as-deposited ZnO nanostructure sample shows the lowest sensitivity and linearity as pH sensing membrane. The post annealing at 300 °C cause the improvement of sensitivity and linearity from 3.4 mV/pH, 0.0683 to 39.8 mV/pH, 0.9672. The lowest sensitivity of as-deposited sample may due to the low adhesion of ZnO nanorods with ITO substrate. The fabricated as-deposited samples were powder-like structures that make it easy to peel-off when immersed in pH buffer solutions. The improvement of the sensor performance may be due to the improvement of the crystalline quality. However, further improvement of the crystalline quality by the higher annealing temperature reduce the sensitivity and linearity of sample due to the reduction of binding sites in the membrane [32]. Binding sites are needed to react with H+ ions. According to the site binding theory, the number of binding sites residing on the sensing membrane could lead to changes in the surface potential voltage between the sensing layer and the electrolyte interface. The actual change of the surface potential voltage would depend on the pH value of the electrolytic solution. Based on the above-mentioned site binding model, the surface potential voltage ($\psi 0$) between the sensing layer and the electrolytic interface can be expressed as [4]:

$$2.303\ (pHpzc - pH) = \frac{q\psi 0}{kT} + sinh^{-1}\left(\frac{q\psi 0}{kT} - \frac{1}{\beta}\right) \quad (1)$$

where pH_{pzc} is the pH value at the point of zero charge, q is the electron charge, k is the Boltzmann's constant, T is the absolute temperature, and β is the sensitivity parameter. The relation between β and the surface sites per unit area (NS) can be expressed as

$$\beta = \frac{2q^2 N_S (K_a K_b)^{1/2}}{KTC_{DL}} \quad (2)$$

where K_a and K_b are acid equilibrium constant and basic equilibrium constant, respectively, and C_{DL} is the capacitance of the electrical double layer derived by the Gouy-Chapman-Stern model. From the equation (1) and (2) it is clear that the sensitivity and the binding sites at the sensing membrane surface have obvious linear response, hence, this result can be attributed to a reduction in the hydride ion reaction site density (N_S) by the raised temperature.

4. Conclusions

The ZnO nanostructures was successfully deposited on ITO substrate using simple CBD process in water bath at 95 °C for 4 h. The post-deposition heat treatment on ZnO nanostructures samples was varied from 200 to 600 °C in order to study the effect of annealing temperature on ZnO nanostrucutres as sensing membrane. From the FESEM images, the as-deposited sample was nanorods structure. Applying the post heat treatment caused the ZnO nanorods to agglomerate on each other and formed the ZnO nanoflowers. The diameter of ZnO nanoflowers increased with the increased of annealing temperature. The XRD results showed that the intensity of ZnO (100) was increased when temperature was increased up to 500 °C before started to decrease at 600 °C. The as-deposited sample sensitivity is only 3.9 mV/pH with linearity of 0.0683. The sensitivity and linearity of the samples increased when the post heat treatment was applied. The highest sensitivity and linearity was obtained from the sample annealed at 300 °C, with 39.9 mV/pH sensitivity and 0.9792 linearity. The post annealed treatment improved the crystallite size which provided high surface area for sensing activity. From the site bonding model also, it can be concluded that the high temperature may induce the number of binding sites residing to reduce and thus the performance of ZnO nanostructures as EGFET sensing membrane decreased.

Acknowledgements

This work is partially supported by the Ministry of Higher Education Malaysia under the FRGS (Project Code: 600-RMI/FRGS 5/3 (0106/2016)). Authors would also like to thank the Institute of Science (IOS), UiTM and Institute of Research Management and Innovation (IRMI), UiTM for their support.

References

[1] O. Gubanova, M. Andrianova, M. Saveliev, N. Komarova, E. Kuznetsov, and A. Kuznetsov, "Fabrication and package of ISFET biosensor for micro volume analysis with the use of direct ink writing approach," *Materials Science in Semiconductor Processing,* vol. 60, pp. 71-78, 2017.

[2] Y. Rayanasukha, S. Pratontep, S. Porntheeraphat, W. Bunjongpru, and J. Nukeaw, "Non-enzymatic urea sensor using molecularly imprinted polymers surface modified based-on ion-sensitive field effect transistor (ISFET)," *Surface and Coatings Technology,* vol. 306, Part A, pp. 147-150, 2016.

[3] S. E. Naimi, B. Hajji, I. Humenyuk, J. Launay, and P. Temple-Boyer, "Temperature influence on pH-ISFET sensor operating in weak and moderate inversion regime: Model and circuitry," *Sensors and Actuators B: Chemical,* vol. 202, pp. 1019-1027, 10/31/ 2014.

[4] K. A. Yusof, R. Abdul Rahman, M. A. Zulkefle, S. H. Herman, and W. F. H. Abdullah, "EGFET pH Sensor Performance Dependence on Sputtered TiO_2 Sensing Membrane Deposition Temperature," *Journal of Sensors,* vol. 2016, p. 9, 2016.

[5] F. Lin, H. Y. Chang, S. H. Hsiao, H. I. Chen, and W. C. Liu, "Preparation and characterization of nickel oxide-based EGFET pH sensors," in *2015 9th International Conference on Sensing Technology (ICST)*, 2015, pp. 402-405.
[6] A. Das, D. H. Ko, C.-H. Chen, L.-B. Chang, C.-S. Lai, F.-C. Chu, *et al.*, "Highly sensitive palladium oxide thin film extended gate FETs as pH sensor," *Sensors and Actuators B: Chemical,* vol. 205, pp. 199-205, 2014.
[7] H. Guliga, W. F. H. Abdullah, and S. H. Herman, "Extended gate field effect transistor (EGFET) integrated readout interfacing circuit for pH sensing," in *2014 2nd International Conference on Electrical, Electronics and System Engineering (ICEESE)*, 2014, pp. 11-14.
[8] C. Hui-Hsin, L. Yi-Ting, Y. Chai-Ming, L. Ji-dung, C. Chiuan-Chian, and L. Chao-Sung, "Characterization of body effect of Au-EGFET for KRAS gene detection," in *2014 IEEE International Nanoelectronics Conference (INEC)*, 2014, pp. 1-4.
[9] C.-E. Lue, I. S. Wang, C.-H. Huang, Y.-T. Shiao, H.-C. Wang, C.-M. Yang, *et al.*, "pH sensing reliability of flexible ITO/PET electrodes on EGFETs prepared by a roll-to-roll process," *Microelectronics Reliability,* vol. 52, pp. 1651-1654, 2012.
[10] L.-T. Yin, Y.-T. Lin, Y.-C. Leu, and C.-Y. Hu, "Enzyme immobilization on nitrocellulose film for pH-EGFET type biosensors," *Sensors and Actuators B: Chemical,* vol. 148, pp. 207-213, 6/30/ 2010.
[11] J.-L. Wang, P.-Y. Yang, T.-Y. Hsieh, C.-C. Hwang, and M.-H. Juang, "pH-Sensing Characteristics of Hydrothermal Al-Doped ZnO Nanostructures," *Journal of Nanomaterials,* vol. 2013, p. 7, 2013.
[12] H.-H. Li, C.-E. Yang, C.-C. Kei, C.-Y. Su, W.-S. Dai, J.-K. Tseng, *et al.*, "Coaxial-structured ZnO/silicon nanowires extended-gate field-effect transistor as pH sensor," *Thin Solid Films,* vol. 529, pp. 173-176, 2013.
[13] C. Schultealbert, T. Baur, A. Schütze, S. Böttcher, and T. Sauerwald, "A novel approach towards calibrated measurement of trace gases using metal oxide semiconductor sensors," *Sensors and Actuators B: Chemical,* vol. 239, pp. 390-396, 2017.
[14] J. Choi, S. Kim, H. Kwon, M. Kim, H. Moon, J. Park, *et al.*, "Organic vapor-jet-based route for solvent-free additive formation of oxide semiconductors," *Organic Electronics,* vol. 43, pp. 235-239, 2017.
[15] J. Sukunta, A. Wisitsoraat, A. Tuantranont, S. Phanichphant, and C. Liewhiran, "Highly-sensitive H2S sensors based on flame-made V-substituted SnO2 sensing films," *Sensors and Actuators B: Chemical,* vol. 242, pp. 1095-1107, 2017.
[16] L. Huo, X. Yang, Z. Liu, X. Tian, T. Qi, X. Wang, *et al.*, "Modulation of potential barrier heights in Co3O4/SnO2 heterojunctions for highly H2-selective sensors," *Sensors and Actuators B: Chemical,* vol. 244, pp. 694-700, 2017.
[17] G. Wang, J. Wang, Y. An, and C. Wang, "Anodization fabrication of 3D TiO2 photonic crystals and their application for chemical sensors," *Superlattices and Microstructures,* vol. 100, pp. 290-295, 2016.
[18] M. T. Man, J.-H. Kim, M. S. Jeong, A.-T. T. Do, and H. S. Lee, "Oriented ZnO nanostructures and their application in photocatalysis," *Journal of Luminescence,* vol. 185, pp. 17-22, 2017.
[19] N. Srinivasan, M. Revathi, and P. Pachamuthu, "Surface and optical properties of undoped and Cu doped ZnO nanostructures," *Optik - International Journal for Light and Electron Optics,* vol. 130, pp. 422-426, 2017.
[20] A. Hassanpour, N. Bogdan, J. A. Capobianco, and P. Bianucci, "Hydrothermal selective growth of low aspect ratio isolated ZnO nanorods," *Materials & Design,* vol. 119, pp. 464-469, 2017.
[21] M. Ahmad, M. A. Iqbal, J. Kiely, R. Luxton, and M. Jabeen, "Enhanced output voltage generation via ZnO nanowires (50 nm): Effect of diameter thinning on voltage enhancement," *Journal of Physics and Chemistry of Solids,* vol. 104, pp. 281-285, 2017.

[22] Y. Jiang, R. Sun, H.-B. Zhang, P. Min, D. Yang, and Z.-Z. Yu, "Graphene-coated ZnO tetrapod whiskers for thermally and electrically conductive epoxy composites," *Composites Part A: Applied Science and Manufacturing,* vol. 94, pp. 104-112, 2017.

[23] Q. J. Feng, J. Wang, S. Liu, R. Z. Xu, K. Tang, Y. Liu*, et al.*, "Fabrication and characterization of highly ordered phosphorus-doped ZnO nanocombs," *Journal of Physics and Chemistry of Solids,* vol. 74, pp. 476-479, 2013.

[24] N. D. H. A. Patah. A. B. Rosli, M. A. Rosdan, M. M. Marbie, M. H. Juhari, S. S. Shariffudin, S. H. Herman, M. Rusop, "Influence of metal catalyst for zinc oxide nanostructures grown by TCVD method for extended-gate FET sensor application," in *Micro and Nanoelectronics (RSM), IEEE Regional Symposium on*, 2013, pp. 171-174.

[25] K. G. Girija, K. Somasundaram, A. Topkar, and R. K. Vatsa, "Highly selective H2S gas sensor based on Cu-doped ZnO nanocrystalline films deposited by RF magnetron sputtering of powder target," *Journal of Alloys and Compounds,* vol. 684, pp. 15-20, 2016.

[26] A. B. Rosli, N. D. H. A. Patah, M. A. Rosdan, M. M. Marbie, M. H. Juhari, S. S. Shariffudin, *et al.*, "Influence of metal catalyst for zinc oxide nanostructures grown by TCVD method for extended-gate FET sensor application," in *Micro and Nanoelectronics (RSM), 2013 IEEE Regional Symposium on*, 2013, pp. 171-174.

[27] G. M. Ali, R. H. Dhaher, and A. A. Abdullateef, "pH sensing characteristics of EGFET based on Pd-doped ZnO thin films synthesized by sol-gel method," in *2015 Third International Conference on Technological Advances in Electrical, Electronics and Computer Engineering (TAEECE)*, 2015, pp. 234-238.

[28] M. Stefan, D. Ghica, S. V. Nistor, A. V. Maraloiu, and R. Plugaru, "Mn2+ ions distribution in doped sol–gel deposited ZnO films," *Applied Surface Science,* vol. 396, pp. 1880-1889, 2017.

[29] L. C.-K. Liau and J.-S. Huang, "Energy-level variations of Cu-doped ZnO fabricated through sol-gel processing," *Journal of Alloys and Compounds,* vol. 702, pp. 153-160, 2017.

[30] M. A. Zulkefle, R. A. Rahman, M. R. Mahmood, W. F. H. Abdullah, and S. H. Herman, "Sensing capability of TiO2 thin films with different thicknesses as sensing membrane of EGFET pH sensor," in *2015 International Symposium on Technology Management and Emerging Technologies (ISTMET)*, 2015, pp. 30-33.

[31] A. B. Rosli, M. M. Marbie, S. H. Herman, and M. H. Ani, "Gold-Catalyzed Growth of Aluminium-Doped Zinc Oxide Nanorods by Sputtering Method," *Journal of Nanomaterials,* vol. 2014, p. 7, 2014.

[32] P. C. Yao, M. C. Lee, and J. L. Chiang, "Annealing Effect of Sol-Gel TiO_2 Thin Film on pH-EGFET Sensor," in *2014 International Symposium on Computer, Consumer and Control*, 2014, pp. 577-580.

Optical Fiber Thermometer Based on Fiber Bragg Gratings

Ekbal Bin Rosli, Uzer Mohd. Noor

Bachelor of Engineering (Hons) Electronic Engineering, Faculty of Electrical Engineering, Universiti Teknologi Mara Shah Alam 40450 Selangor, Malaysia

E-mail: ekbal.rosli@yahoo.com

Abstract
Fiber Bragg grating has generated much interest in use as sensors to measure strain, temperature, and other physical parameters. It also the most common component used to develop this sensor with the advantages of simple, intrinsic sensing elements, electrically passive operation, EMI immunity, high sensitivity, compact size and potentially low cost [6]. This paper reports the design of an optical fiber thermometer based on fiber Bragg gratings. The system was developed for detecting temperature and strain by monitoring the shift of Bragg wavelength. The shifting of Bragg wavelength is used to indicate the temperature and strain due to the change in the surrounding temperature and strain. When the temperature and strain reach the exact wavelength level of the system, the temperature and strain value will display on the Arduino liquid crystal display (LCD). The optical fiber will provide the broadband light source and after passing the FBG the Bragg wavelength into the optical spectrum analyzer (OSA). The system is based on FBG as a physical quantity sensor. The temperatures measured is taken from the water bath and that of the strain is provided by amount of slotted mass used. The outcome of this project is to characterize the Bragg wavelength shifting from the fiber Bragg grating output. As the conclusion, this project provides an efficient optical fiber thermometer in measuring temperature and strain in order to replace the use of conventional electrical instruments.

Keywords— Fiber Bragg grating (FBG), Optical Spectrum Analyzer (OSA)

1. Introduction

The communications industry has developed rapidly over the past two decades from an orderly and steadily growing enterprise to a chaotic marketplace with continually altered regulations, complex business relationship and explosive optical technology leading to dynamic growth. The installation of optical fibers, which is directly linked to telecommunications or sensors, have grown consistently this decade and it cannot be denied anymore. Just to make sure we can stand with the others in this growth of technology, a research and innovation in this topic need to be done as soon as possible.

Fiber Bragg gratings use light as the medium to send the data and have low attenuation with high security compared to the conventional temperature and strain sensor that can easily have power loss and can be tapped from outside. It is safer and can withstand high temperature and accurate in measuring the temperature and strain. Fiber Bragg grating works by reflecting a specified light wavelength but transmitted the other wavelengths to the detector. When the FBG is heated, the central wavelength will experience a red shift but at room temperature, it will return to its original wavelength instantly [2]. Based on the Fresnel reflection, by varying the refractive index or physical dimensions of the waveguiding core the different periodic modulation of the refractive index can be obtained. This feature varies the reflected light modulation and the reflected wavelength is called

Bragg-wavelength,λ_B. Wavelength shifting is the main element used in optical fiber Bragg gratings sensor. This shifting affected by manipulating the measurand (e.g. temperature and strain) apply on the fiber Bragg grating. When light guided along the core of the optical fiber, it will be scattered by each grating line.

Finally, suppose the theoretical calculation are good enough to characterize the Fiber Bragg grating, but since many manufacturers do not want to share the fiber Bragg gratings characteristics due to company secrecy or to avoid any competition from other company. So that, the experiment conducted to characterize the Fiber Bragg grating need to be done.

2. METHODOLOGY

Along with this project, a lot of precaution and method need to be followed to complete the project. Before that, the flow chart of the project needs to design to get the exact ways to develop the research. The project starts with research on the Fiber Bragg Grating and Arduino programming technique. The understanding of theory, characteristics, and properties of Fiber Bragg grating need to be studied. Arduino is a system used to design a device to record the temperature of the waterbath. This device is using C programming language to process the data in the microcontroller and send the command throughout the system. It is widely used in the electronic world to interface with the other system with the support of visual basic 6 software.

The second step is to set up the experiment for the characterization of Fiber Bragg grating with the temperature or strain. The Fiber Bragg grating will be placed in the water bath temperature to study the behavior of the fiber Bragg gratings (FBG). The water will be used as the medium to transfer heat onto the FBG. A normal water has the temperature between 0°C up to 100°C in Celsius or 32°F up to 212°F in Fahrenheit. However, to increase the boiling point and decrease the freezing point, the adding of 250g sodium chloride (NaCl) is used to change the water characteristic. By doing this new freezing point is produce which is -10°C and the boiling point is 110°C. To characterize the Fiber Bragg grating, the water bath temperature will be set from -10°C up to 60°C with 10°C range, then the value of Bragg wavelength will be recorded in the optical spectrum analyser (OSA). The data collected was transferred into the Microsoft Excel to be analyzed and plotted a graph. In addition, the small range of temperature also been recorded from 0°C up to 10°C with 2°C increments. Next, for the strain, the experiment will be tested with the copper load and the mass of 0g up to 30g with 5g increment was been used. Same the data will be analyzed in Microsoft Excel.

The data from Bragg wavelength against temperature or strain will be used as a reference to decide which Fiber Bragg grating will be more suitable to be used as optical fiber thermometer sensor. There are 4 selections for this experiment, four bare fiber Bragg grating with different center wavelength will be used. By creating a straight-line graph, the best optical fiber thermometer can be choose. The smoother the graph straight-line, the sensitive the Fiber Bragg Grating. Lastly, to create a system that can be a monitor on the place, all the instrument use need to interface with the computer. For this project, there are 2 instruments that need to be interfaced, which is optical spectrum analyzer (OSA) and Arduino Mega 2560. The system uses Arduino Mega 2560 as a base system to control the system to display the output temperature of the waterbath. As the waterbath temperature changes the Bragg wavelength will be changed as well and the output data sent to Arduino to be processed. The changing of Bragg wavelength will be analysed.

Design selection need be done first by listing the hardware and software choice that most suitable for this purpose optical fiber thermometer. For hardware design, it is divided into four

IOP Conf. Series: Materials Science and Engineering **340** (2018) 012020 doi:10.1088/1757-899X/340/1/012020

categories which are Fiber Bragg grating, Power supply, Wavelength Analyzer and Materials for medium tested. There are many types of fiber Bragg grating used in the industries because Fiber Bragg grating is a multi-purpose sensor and quite sensitive to the stimulus. It also saves from any interference. Since this system is focused on the optical fiber thermometer, two type of fiber Bragg grating being selected, which bare Fiber Bragg grating and shield Fiber Bragg gratings. Bare fiber Bragg grating has been used because it is more sensitive and directly contact to the material tested compare to the shielded that may slightly change the temperature based on shielded materials used. Next, to turn on the Arduino system, it needs a power supply. Two choices already are chosen either to use triple output DC power supply or AC to DC adapter. For this project, AC to DC adapter is used because it has fixed voltage and current for the Arduino microcontroller. In addition, the supply used only for the Arduino board. So, AC to DC adapter was most suitable to be used. For Bragg wavelength analyzer, laboratory equipment is needed to be used. Only optical spectrum analyzzer are being selected for wavelength reader because of the waveform display and more precise to collect the data.

Moreover in this project, two types of the testing to be done which are temperature and strain. Firstly, the temperature is divided into two conditions which are hot air and water bath. The reason water bath has been chosen because of the ability of water that easier to control either in hot condition or cold condition. The characteristic of the water can be changed by adding the impurities in it. It also can change the boiling point and the freezing point better than the hot air that hard to control. The results may be varied due to a different level of air temperature. So, for this project, the water bath is suitable to see the Bragg wavelength change when the fiber Bragg grating makes a contact with the medium tested. Secondly, for the strain, the choices are divided into two which are fixed copper load and plasticine. Plasticine has an ability to change its mass by weighing it on the weighing balance. But for this experiment, fixed copper load is more suitable to be used due to firm shape compare to plasticine that may slightly change in mass.

Move on to software, it is divided into three categories which are LabVIEW, Agilent VEE Pro, and Arduino programming. This three software commonly being used to interface between computer and instrument. For this project, Arduino programming is being selected to process the data in the Arduino Mega 2560 microcontroller.

3. RESULT AND DISCUSSION

The optical fiber thermometer prototype is setup as in fig. 1. The comparison of the four fiber Bragg grating is tested to see the behaviour of the FBG. To characterize fiber Bragg grating and water bath temperature of an experiment, the instrument is arranged as shown in fig. 1. The results of the eight temperature for four different FBG were stated in Table I.

a. Characterization of FBG in every 10°C increments (large range).

To record the water bath temperature, 18B20 temperature sensor (waterproof) is used. While running the temperature sensor, fiber Bragg grating was immersed together with the temperature sensor in the water bath. So, that the real temperature can be measured. The used of IR thermometer sensor is used to make sure the temperature used exactly as in the objective.

In order to control the temperature, hot plate machine have been used as a heating element to maintain the temperature. The hot plate temperature will be adjusted so that the water temperature will be maintained at certain degree Celsius. The data of the Bragg waveform have been collected and the center wavelength has been used to construct the graph to see the sensitivity of the fiber Bragg behaviour. The graph for four FBG results has been constructed as shown in fig. 2. The linear graph has been formed and proves the theory of the fiber Bragg gratings as the temperature increased, the Bragg wavelength also increased. Means the shifting occurs on the Bragg grating of the FBG when the temperature applied on it. With this results, the project can be carried out to make the fiber Bragg gratings as an optical fiber thermometer.

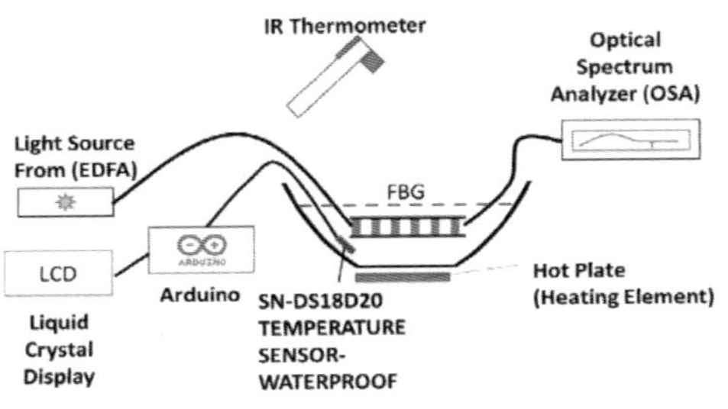

Figure 1. Set-up for characterization of fiber Bragg grating to measure temperature

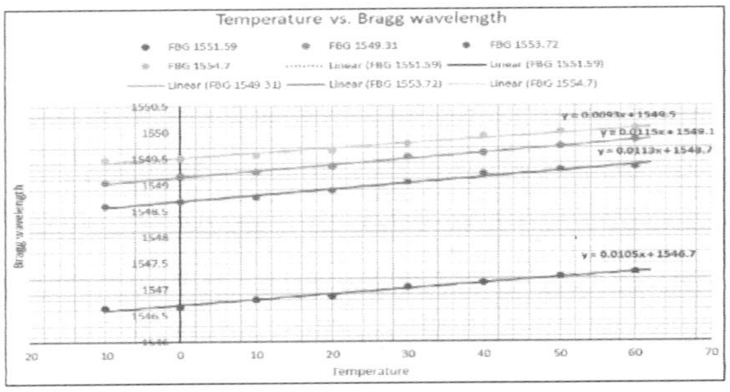

Figure 2. Plot of temperature against Bragg wavelength

TABLE I. THE COMPARISON BETWEEN WATER BATH TEMPERATURE (LARGE RANGE) WITH DIFFERENT TYPES OF FIBER BRAGG GRATING.

Temp(°C)	Types of Fiber Bragg gratings			
	FBG 1553.72	FBG 1554.7	FBG 1549.31	FBG 1551.59
-10	1548.60	1549.48	1549.04	1546.64
0	1548.68	1549.50	1549.16	1546.66
10	1548.76	1549.56	1549.24	1546.80
20	1548.88	1549.64	1549.34	1546.86
30	1549.04	1549.78	1549.52	1547.04
40	1549.20	1549.92	1549.60	1547.12
50	1549.28	1550.00	1549.72	1547.24
60	1549.32	1550.06	1549.84	1547.32

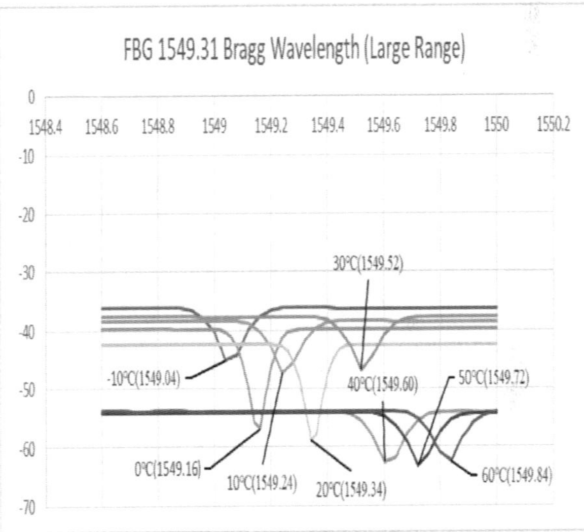

Figure 3. Waveform for FBG 1549.31 at different temperature

Based on the graph plot from fig. 2, the graph is linear, and it proves that the bare fiber Bragg grating (FBG 1549.31) have the good efficiency. The straight line of the graph is y=0.0115x + 1549.1.

From fig. 3, shows the shifting of the Bragg wavelength with different temperature. Among four FBG, only FBG 1549.31 is the best FBG to be used as an optical fiber thermometer in large range due to high sensitivity compared to other FBG. The different between the temperature tested on the FBG 1549.31 has a small different and almost constant with each other. The shifting of the Bragg wavelength move from left to right and this proves another theory as mention in the introduction part.

b. Characterization of FBG in every 2°C increments (small range).

Move on to small measurement to see the sensitivity of the FBG with the small range of the temperature. Six temperature has already been used to see the consistency of Bragg wavelength produced.

TABLE II. THE COMPARISON BETWEEN WATER BATH TEMPERATURE (SMALL RANGE) WITH DIFFERENT TYPES OF FIBER BRAGG GRATING.

Temp(°C)	Types of Fiber Bragg gratings			
	FBG 1553.72	FBG 1554.7	FBG 1549.31	FBG 1551.59
0	1548.68	1549.50	1549.16	1546.66
2	1548.72	1549.52	1549.16	1546.72
4	1548.72	1549.54	1549.20	1546.72
6	1548.76	1549.54	1549.24	1546.76
8	1548.76	1549.56	1549.24	1546.76
10	1548.76	1549.56	1549.28	1546.80

TABLE III. THE COMPARISON BETWEEN MASS) AND BRAGG WAVELENGTH OF BARE FIBER BRAGG WAVELENGTH.

Mass (g)	Types of Fiber Bragg gratings			
	FBG 1553.72	FBG 1554.7	FBG 1549.31	FBG 1551.59
0	1549.24	1549.68	1549.44	1546.88
5	1549.28	1549.69	1549.48	1546.96
10	1549.32	1549.69	1549.52	1547.00
15	1549.32	1549.73	1549.56	1547.04
20	1549.36	1549.73	1549.60	1547.08
25	1549.36	1549.76	1549.64	1547.08
30	1549.40	1549.76	1549.68	1547.12

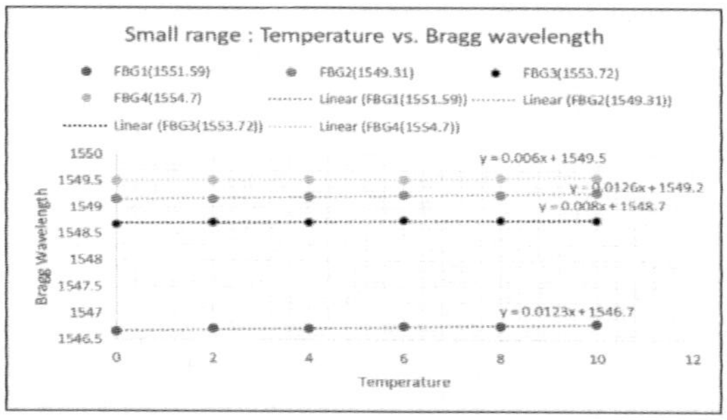

Figure 4. Plot Of The Temperature Against Bragg Wavelength (Small Range)

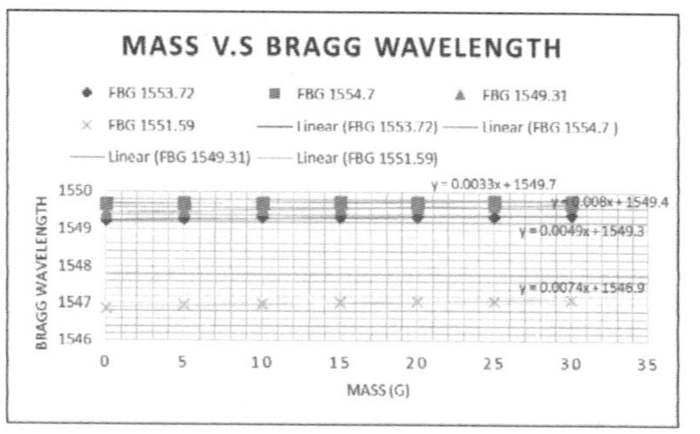

Figure 5. Plot of mass against Bragg wavelength

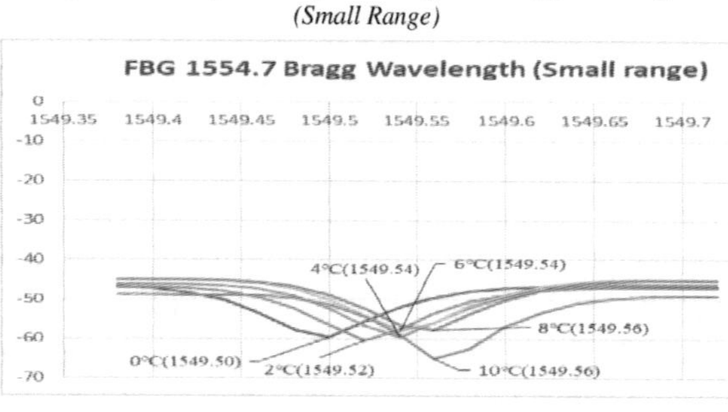

Figure 6. Waveform for FBG 1554.7 at different temperature

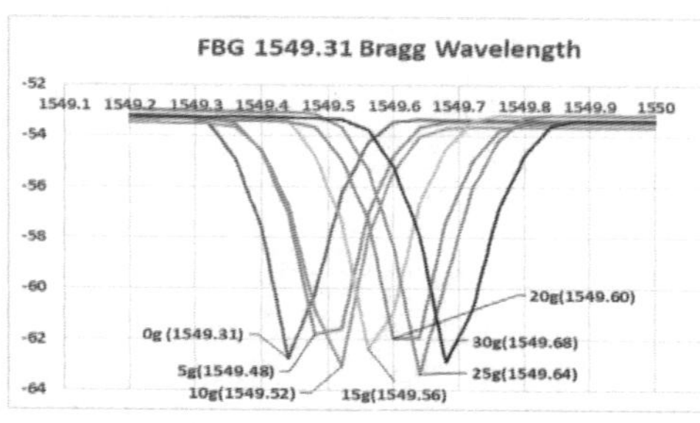

Figure 7. Waveform of FBG 1549.31 at different mass

From the graph above, the result shows the linear graph of the Bragg wavelength increased but only FBG 1554.7 manage to show the sensitivity of the FBG within small range value and the waveform shown as in fig. 6.

c. Characterization of FBG in every 5g increment of mass.

Seven different mass have been used to see the Bragg wavelength changes on the four different types of FBG and the graph has been plotted as shown in fig. 6.The graph above shows the linear graph of the increment of the Bragg wavelength due to the elongation of the FBG in the stimulus of the mass added to the system. As the mass increase, the length on the FBG also increases cause the change of refractive index in the fiber.

Lastly, from the graph plotted in fig. 6, the deduction for the best FBG for strain can be concluded. FBG 1549.31 already been chosen because the sensitivity and the waveform produced on the OSA. The graph result as shown in fig. 7 shown the strain graph to support the theory research. The shifting of the Bragg wavelength to the right gives the picture of the elongation of the FBG due to the mass applied on it. The length of the grating to another grating is increasing and the refractive index in the FBG also changing. The author still trying to interface the Arduino with the optical spectrum analyzer and the research still continues to make the system connected between OSA and Arduino.

4. CONCLUSION

Based on the results obtained, there is a good linear relationship between temperature applied on FBG to the Bragg wavelength shift using FBG as an optical fiber thermometer. The sensitivity of wavelength shift to the changing of temperature is suitable for FBG to monitor the temperature because of its precision on the temperature change. The issue of strain effect on the FBG sensitivity has been solved. Bare fiber Bragg grating sensor is the most suitable approach to replace the conventional electrical sensor in measuring temperature and strain of the substances. The convenience of fiber Bragg grating gives a lot of benefits to be used in the various environment.

For future recommendation, the author like to recommend the testing of fiber Bragg grating at high temperature and strain because of the unique characteristic of the optical fiber. Further enhancement in optical fiber sensor may expand the use of the fiber optic and replace the conventional electrical sensor in future. Plus, reduce the use of metal in the sensor appliances.

5. REFERENCES

[1] Chehura, E.; James, S.W.; Tatam, R. Temperature and strain discrimination using a single tilted fiber Bragg grating. Opt. Commun., pp. 344–347, 2007
[2] Yue-hong Zhu; Hai Qi; Guang-xiang Wang; "Temperature characteristic of fiber Bragg grating," Electronic and Mechanical Engineering and Information Technology (EMEIT), 2011 International Conference on, vol.8, no., pp.4020-4022, 12-14 Aug. 2011
[3] W. B. S. J. Eric Udd, Fiber Optic Sensors: An Introduction for Engineers and Scientists, New Jersey: John Wiley & Sons, Inc., 2011.
[4] P. B. R., T. S. Y. Shizhuo Yin, Fiber optic sensor, Broken Sound Parkway New York: CRC Press Taylor & Francis Group, 2008.
[5] K. K. Andreas Othonos, Fiber Bragg Gratings: Fundamental and Applications in Telecommunications and Sensing, Boston, London: Artech House Publishers, pp. 4309-4341, 1997.
[6] Xinyoung Dong, Y. L. Simultaneous displacement and temperature measurement with cantilever-based fiber Bragg grating sensor. NH Elsevier, pp. 1-5, 2001

6. ACKNOWLEDGMENT

The authors would like to appreciate the fullest support of the Faculty of Electrical Engineering, Universiti Teknologi Mara Shah Alam, Selangor in completing the project successfully.

Institute of Physics
Dirac House, Temple Back
Bristol BS1 6BE UK

ISSN: 1757-8981
ISBN 978-1-5108-6116-9
9 781510 861169

CURRAN ASSOCIATES INC.
proceedings.com

26th International Laser Physics Workshop (LPHYS'17)

Journal of Physics: Conference Series Volume 999

Kazan, Russia
17 – 21 July 2017

IOP
Institute of Physics